Schötz

Sampling of Injection Molds

Andreas Schötz

Sampling of Injection Molds

Structured and Analytical Process

HANSER

FSC
www.fsc.org
MIX
Paper from responsible sources
FSC® C083411

Print-ISBN: 978-1-56990-196-0
E-Book-ISBN: 978-1-56990-300-1

Bibliographic information of the German National Library:
The German National Library lists this publication in the German National Bibliography; detailed bibliographic data are available on the Internet at http://dnb.d-nb.de.

www.hanserpublications.com
www.hanser-fachbuch.de
Editor: Dr. Mark Smith
Production Management: Cornelia Speckmaier
Cover concept: Marc Müller-Bremer, *www.rebranding.de*, Munich
Cover design: Tom West
Cover picture: © Andreas Schötz
Typesetting: le-tex publishing services GmbH, Leipzig
Printed and bound by: CPI Books GmbH, Leck
Printed in Germany

Contents

Preface

The intention behind writing this technical book is that mold sampling is often regarded as a minor matter in the plastics processing industry. This should not be the case, as it is the most important step in the process for achieving a flawless injection mold and molded part. Practical experience shows that many things have a negative influence on optimal mold sampling in the injection molding industry, such as:

- The time pressure in mold sampling, as resources for series production are eliminated
- Deficits in the process knowledge of the employees
- Incorrect procedures for finding the optimum machine parameters
- Systemless "testing" of machine setting parameters in the event of molded part errors and process problems
- Important working steps for of a sampling are forgotten or skipped
- Lack of or inadequate training and further training of employees
- Lack of sampling documentation in the form of templates and checklists
- Poor to partial absence of communication among the employees involved in the sampling process when identifying problems and causes
- Lack of specifications, such as standardization, in the course of mold sampling
- Too many necessary mold-optimization loops during the sampling phase

This book is intended to sensitize the reader to the problems described above and to provide guidance on how to perform optimal, structured and analytical mold sampling in the company. Offering a complete sequence of the individual sampling steps and a lot of background information, hints, practical examples as well as practical tips, the book accompanies the reader from placement of a sampling order up to hand-over to series production.

The main topics covered are a structured procedure for sampling from the aspects of energy efficiency; documentation and communication of a sampling analysis; optimal machine setting through applying strategic procedures and methodologies on the injection molding machine; process optimization followed by an investigation of process capability as well as a Run@Rate process.

This technical book has been designed in such a way that it is ideally suited for the operator at the injection molding machine as well as for teaching purposes in vocational schools, further education institutes and universities.

My main goal is to give the reader valuable impulses and suggestions for optimally implementing sampling–which is, after all, a very complex process—so that it can be implemented more easily.

Andreas Schötz

The Author

Andreas Schötz, Dipl.-Ing. (FH), started his professional career as an apprentice process mechanic for plastics and rubber technology with a focus on molded parts. After successfully completing his training, he continued working for several years as a skilled worker and shift supervisor in the company where he was trained. Even as a young skilled worker, he recognized the possibilities of being able to make a difference in this technical, up-and-coming profession. He laid the foundations for this by undergoing further training to become a state-certified plastics technician, which enabled him to gain further years of practical experience as a technician. He went on to study plastics technology at the Rosenheim University of Applied Sciences. After graduating, Schötz worked as a development and process engineer for a medium-sized supplier in the automotive sector. His focus was on technical consulting, product development, injection mold sampling and process optimization, injection molding simulation and internal employee training. He became self-employed in September 2013.

Ingenieurbüro Schötz Kunststofftechnik GmbH specializes in the following services in the field of injection molding technology:

- Technical advice
- Holistic process optimization
- Sampling/troubleshooting/process evaluations
- Software & digitization with the specially developed IT platform "IMG"
- Trainings, seminars and workshops
- Research and development (R&D)

Ingenieurbüro SCHÖTZ KUNSTSTOFFTECHNIK GmbH

More information at:
www.kunststofftechnik-ingenieurbuero.de

Information on the Book Structure

■ Mold sampling checklist

In the following chapters of this reference book, the optimum sequence for a sampling procedure is described. As an aid and for self-monitoring of the most important sampling steps, a "sampling checklist" has been drawn up for the "specialist for mold sampling".

The "mold sampling checklist" is structured in such a way that controlled work can be carried out in nine blocks. The advantage of the checklist is that routine processes do not have to be carried out again and again, thus minimizing the risk of forgetting important sampling steps. This checklist will accompany the "mold sampling specialist" through the entire sampling process.

Sampling checklist		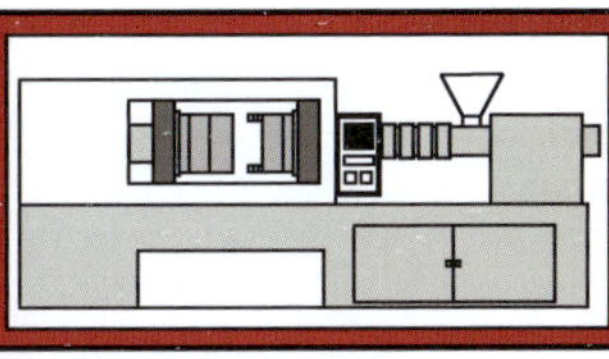
Article description:		**Type of sampling:**
Article number:		☐ First of Tool (FOT)
Mold number:		☐ Subsequent sampling
Customer:		☐ Material sampling
Date of mold sampling:		☐ Complaint
Machine number:		☐ Other
Mold sampling specialist:		
Please document all deviations (fields ticked with "NO") in the sampling report with reasons/comments.		

Gathering information and preparing for sampling				
1	Information sheet for sampling available and completed in full?			Block 1
2	Preparation for sampling on the injection molding machine completed?			

Mold set-up				
Before fitting the mold in the machine:		**Yes**	**No**	
3	Safety checks carried out on the injection molding machine (crane or peripherals) and no defects found?			
4	**Checking the injection mold:**			
4.1	Mold transport bridge present and mounted?			
4.2	Are the labels on the injection molds present and visible?			
4.3	Are there devices for placing the injection molding tool on the floor (optional)?			
4.4	Mold free of visual damage?			
4.5	Is the jaw or slider movement (locking) limited?			
4.6	Ejector plate safety device available?			Block 2
4.7	Have the mold installation dimensions (mold centring Ø nozzle side), installation height, length of the ejector ram, sufficient machine tie bar distance, etc.) been checked?			
4.8	Cooling circuits continuous (compressed air test)?			
4.9	Are the mold temperature control connections for the inputs and outputs correctly labelled?			
4.10	Flow rate (l/min) checked and documented in the machine setting log?			
For hot runner molds:				
4.11	Functional test of the hot runner system (HKS) performed and OK?			

Mold fitting:		**Yes**	**No**	
5	Eyebolt in the mold center of gravity?			
6	Mold mounting grooves with sufficient clamping surface?			
7	Correct screw-in depth (approx. 1.5 × Ø) of the screws observed?			Block 3
8	Can mold cooling be connected without any problems?			
9	Can the nozzle side and ejector side be removed ?			

Basic setting of the clamping unit				
Mold movements:		**Yes**	**No**	
10	Moving mold parts lubricated?			
11	Moving mold parts (slider, ejector, etc.) running smoothly?			
12	Mold movements (travel paths, speeds, ejectors, additional functions in the mold (cores etc.)) set?			**Block 4**
12.1	Mold zeroed?			
13	Mold and checked for function?			
14	Basic setting of the mold clamping force calculated and set?			
15	Correct mold temperature set?			

Basic setting of the plasticizing unit				
Temperatures:		**Yes**	**No**	
16	Barrel temperature correctly selected and set?			
17	Barrel temperature profile matched to thermoplastic?			
18	Correct flange temperature selected and set?			
Plasticizing process:		**Yes**	**No**	
19	Required plasticizing volume or stroke determined and set?			
20	Plasticizing speed set?			
21	Screw back pressure and screw decompression set?			
Injection process:		**Yes**	**No**	**Block 5**
22	Injection pressure set?			
23	Injection speed and injection speed profile set?			
Cooling time and demolding temperature:		**Yes**	**No**	
24	Required demolding temperature determined?			
25	Theoretical residual cooling time calculated and set?			
Plasticizing unit:		**Yes**	**No**	
26	Nozzle contact point zeroed?			
27	Nozzle impression created and OK?			
28	Movements (e.g. lifting of the nozzle) set?			

Filling study, holding pressure and mold clamping force				
Assessment of the plastic melt:		**Yes**	**No**	
29	Melt temperature and mold wall temperature checked?			
30	Assessment of the melt carried out?			
Filling study, holding pressure and mold clamping force:		**Yes**	**No**	
31	Filling study (up to 100% volumetrically filled parts) produced and holding pressure level optimized?			**Block 6**
32	Filling study evaluated and documented?			
33	Required holding pressure time (sealing time) determined and documented via the molded part weight?			
34	Required clamping force (kN or t) optimized with dial gauge?			

Manufacturing first tooling samples and conducting a sampling analysis of the basic setting				
Producing the first tooling samples:		**Yes**	**No**	
35	Machine capability in accordance with VDMA 24.470 Parts 1 and 2?			
36	Cavity pressure sensors connected to the machine?			
37	Injection molding process run in over several cycles (20–30)?			
Sampling analysis of the basic setting:		**Yes**	**No**	
38	Analysis of the required flow rate of the temperature control medium carried out?			
39	All faults occurring on the molded part localized and defined?			**Block 7**
40	Cause of error on the molded part localized and defined?			
41	Specific analysis of injection molding process and mold carried out?			
42	**General/visual inspection of the process:**			
43	Target/actual comparisons of all machine setting parameters carried out?			
44	Is the existing residual melt cushion OK and documented in the machine setting data sheet?			
45	Is the dwell time of the melt in the barrel calculated, checked and documented in the machine setting data sheet?			

Manufacturing first tooling samples and conducting a sampling analysis of the basic setting				
Producing the first tooling samples:		**Yes**	**No**	
46	**Thermal checks (thermography):**			
47.1	Thermal pictures or videos (thermography) taken?			
47.2	Demolding temperature (target/actual comparison) OK?			
47.3	Temperature distribution in the mold and molded part OK?			
47.4	Temperature difference between flow and return temperature of the temperature control unit measured and OK?			
47.5	Is the cavity temperature homogeneous?			
47.6	No hot spots in the mold or on the molded part?			
48	**Cavity pressure curve:**			
48.1	Cavity pressure curves recorded and checked?			
48.2	Injection phase optimal?			
48.3	Compression phase optimal?			
48.4	Correct switching point for holding pressure?			
48.5	Holding pressure level and holding pressure time optimal?			
48.6	Not demolded under residual pressure?			
49	**Evaluation of the molded part and the mold:**			
49.1	Molded part free of ejector marks?			**Block 7**
49.2	Correct mold cavity codes?			
49.3	Date clock and index set correctly?			
49.4	Molded part free of flash?			
49.5	Molded part free of ejector marks?			
49.6	Ejector not protruding?			
49.7	Ejector not retracted?			
49.8	Molded part fully injected?			
49.9	Are the installed mold vents functional?			
49.10	Are the dimensions of the molded part OK?			
49.11	Molded part without bubbles/air bubbles/air pockets?			
49.12	Molded part without warpage?			
49.13	Molded part without sink marks?			
49.14	Molded part without streaks?			
49.15	Molded part without scratches or marks?			
49.16	Molded part without stress cracks (white crack)?			
49.17	Molded part without shiny spots?			
49.18	Molded part without matt spots?			

Manufacturing first tooling samples and conducting a sampling analysis of the basic setting				
Producing the first tooling samples:		**Yes**	**No**	
49.19	Molded part without flow and weld lines in visible or problem areas?			
49.20	Molded part without contamination?			
49.21	Molded part without jetting?			
49.22	Molded part without cold displacements (material)?			
49.23	Injection molding (sprue area) visually OK?			
49.24	Injection molding (sprue) visually OK?			
49.25	Does the removal device function properly?			**Block 7**
49.26	Molded part without diesel effect/burners?			
49.27	Molded part without vinyl record effect?			
49.28	Demolding of the molded part without problems?			
50	**Evaluation of the sampling analysis:**			
50.1	Was the evaluation carried out and documented in full?			
50.2	Are there no defects in the injection mold?			
50.3	Can the basic setting be optimized without negative effects from the mold?			
50.4	Can mold sampling be continued?			

Optimization of the basic setting				
Optimization of the basic setting–Part 1:		**Yes**	**No**	
51	Optimization strategy defined?			
52	Molding tests carried out?			
53	Molding tests analyzed?			
Optimization of the basic setting–Part 2:		**Yes**	**No**	
54	Basic setting evaluated and optimized for productivity?			
55	Basic setting evaluated and optimized for energy efficiency?			**Block 8**
Optimization of the basic setting–Part 3:		**Yes**	**No**	
56	Process analysis of machine setting and process parameters carried out and evaluated?			
57	Run@Rate process of the optimized basic setting performed and evaluated?			

Documentation of mold sampling				
58	Machine setting and process parameters documented in the machine setting data sheet?			**Block 9**
59	All findings, problems, etc. from the mold inspection documented in detail under "Assessment by the mold sampling specialist" in the mold sampling report?			
60	All documents for the current mold sampling (templates) processed correctly and completely?			
61	Documentation handed over in full to the project manager?			

■ Designations for Mold Sampling Specialists

To make the sampling process more transparent, special symbols have been created in the following table for all specialists involved in sampling. These symbols show in which phase of mold sampling the respective specialist is involved, making it easier to understand the process. In addition, the necessary documents of the individual specialists are also labeled with these symbols.

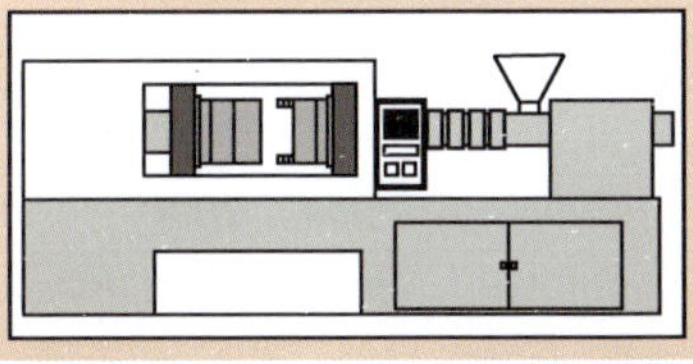	**Mold sampling specialist** E. g. plastics technologist, process engineer
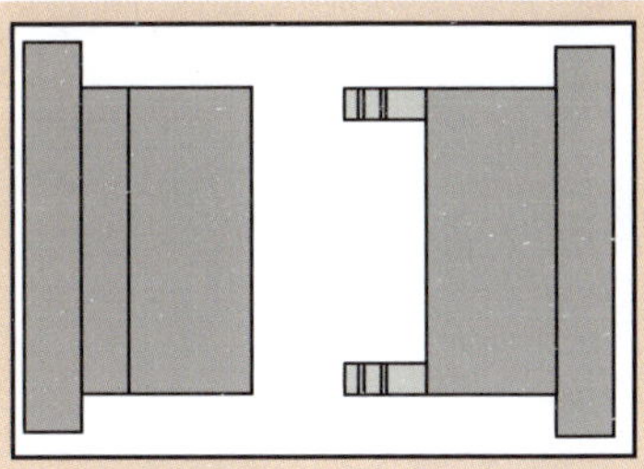	**Mold-making specialist** E. g. mold-maker/tooling shop

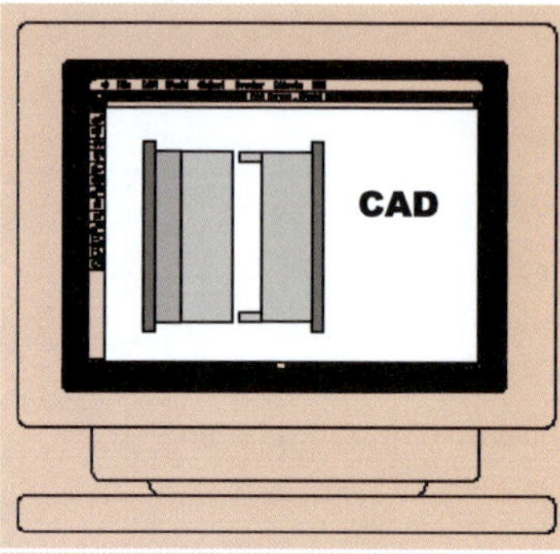

Design specialist
E. g. design engineer

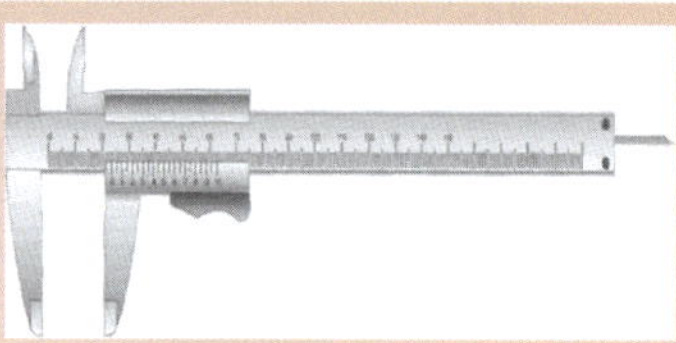

Quality assurance/metrology specialist
E. g. quality manager, quality control

Project manager
E. g. project head, head of department, engineer, technician, supervisor

Description of the Information Boxes

The individual chapters of the textbook are accompanied by information boxes designed to make your work more transparent and clearer.

Practical tip:
Contains valuable practical tips on the respective subject area.

Note:
You should pay particular attention to the following.

Background knowledge:
Background knowledge (further information) on the respective subject area.

Energy efficiency:

Describes where there is great potential for increasing energy efficiency.

Remember:

Important information that you should remember.

Practical example:

Practical examples illustrate the issues described here. The practical example of a "housing cover" will guide you through the book.

1 Introduction

1.1 Why Perform Mold Sampling?

Mold sampling is performed in an injection molding company for every new mold, material change or mold correction. Mold sampling is performed for the following reasons:

- Identifying mechanical defects in the injection mold and rectifying them in a targeted manner
- Strategically and analytically determining, documenting and archiving the process parameters
- Achieving the required molded part quality in terms of appearance and dimensions
- Achieving the optimum cycle time
- Realizing machine-friendly, wear-reduced and energy-efficient series production

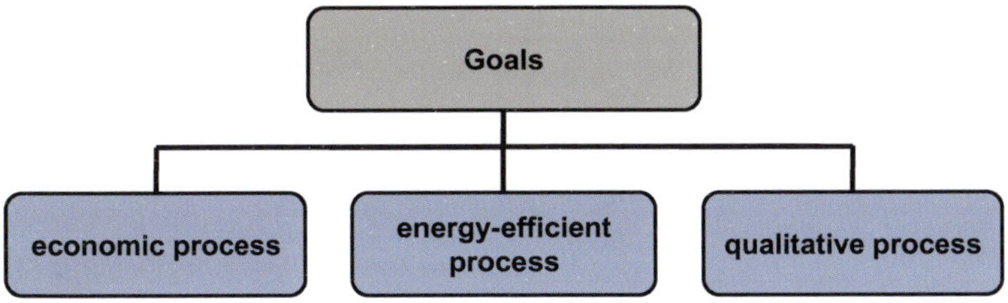

Figure 1.1 Reasons for performing mold sampling

1.2 The Mold Sampling Workflow

Mold sampling is a complex process, as different departments in the company interact with each other. These departments must work together at the right time in order to effectively manage the mold sampling process. This presents every company with a technical and logistical challenge. To make it easier for you to learn and work with this book, the sampling process is clearly illustrated in a flow chart (Figure 1.2) and then systematically covered in the following chapters.

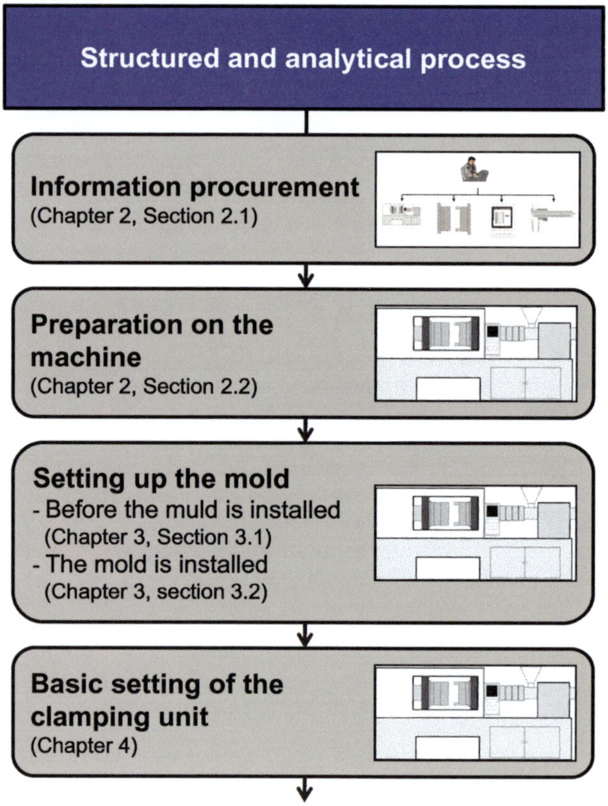

Figure 1.2 Flowchart of the sampling process

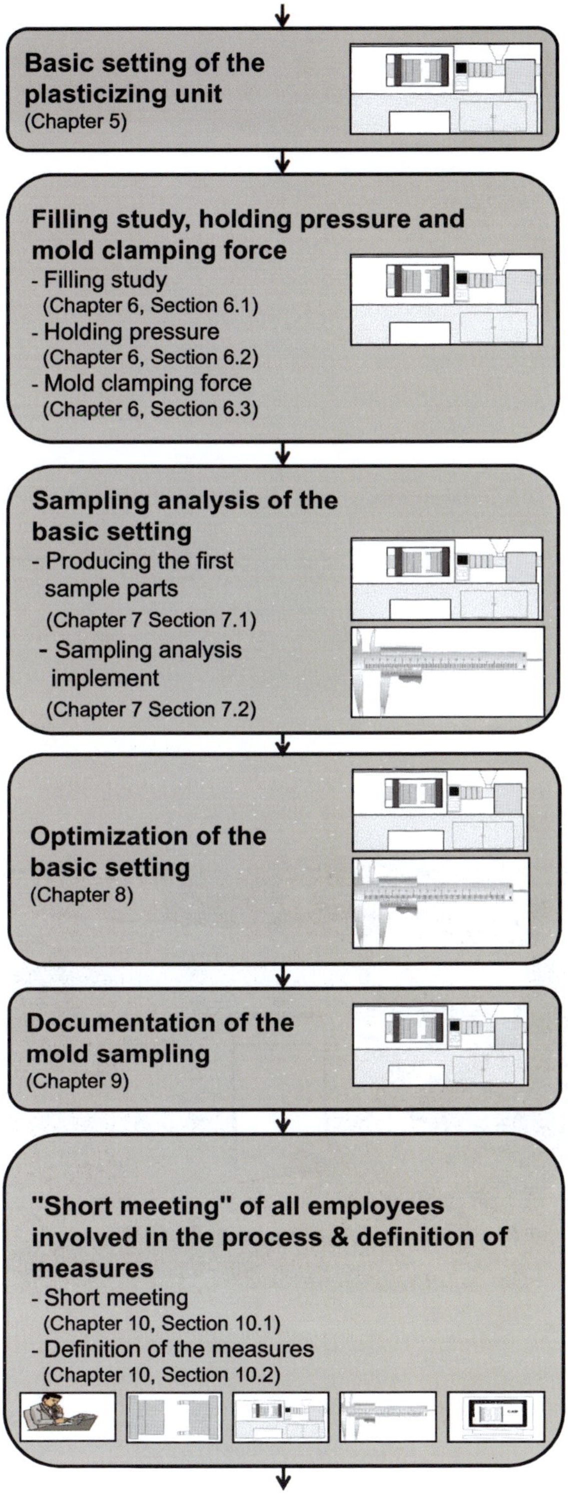

Figure 1.2 Flowchart of the sampling process (*continued*)

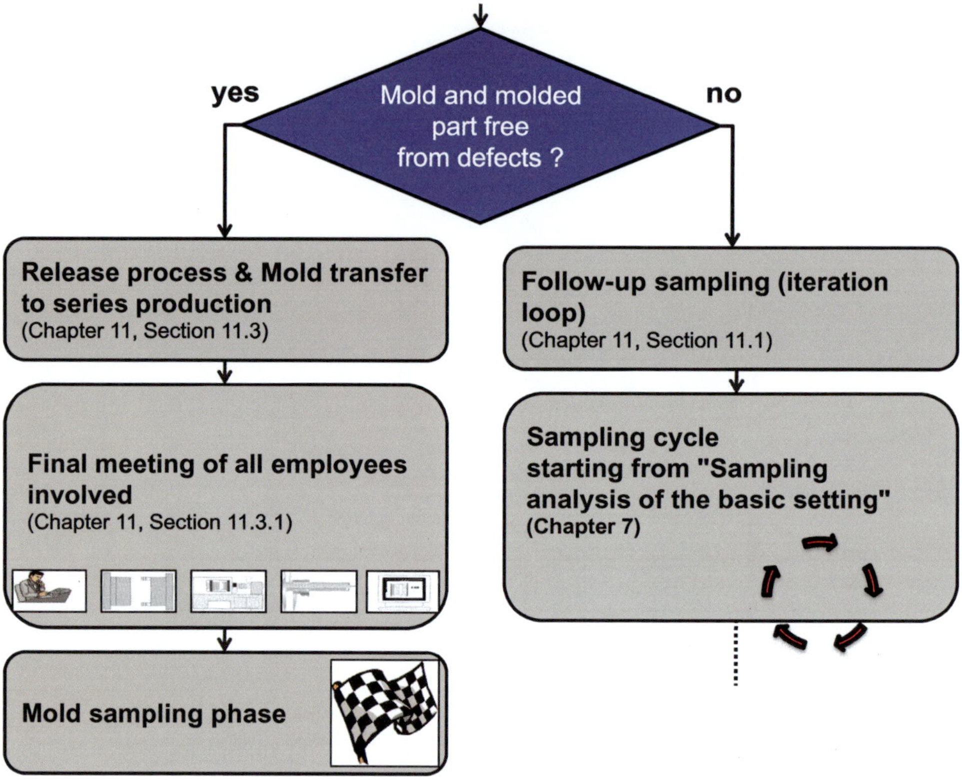

Figure 1.2 Flowchart of the sampling process (*continued*)

1.3 The Time-Factor Problem in the Mold Sampling Process

In an injection molding company, mold sampling ties up the resources of an injection molding machine. This means that series production of injection molded parts cannot take place during the sampling period. As a result, companies often only use a very rough and non-optimal machine setting without mold and process optimization in series production in order to save time in the short term.

The basic attitude of quickly achieving reasonably acceptable and superficial optimization of the process and the mold in order to obtain resources for series production will come back to haunt you later in the form of many problems in series production.

Although this method allows the mold to be handed over from the sampling process to series production more quickly, it can lead to major fluctuations in subsequent series production. This results in higher reject rates and customer complaints, which automatically contribute to higher production costs. This is a basic problem of every sampling process. Consequently, the resources saved in the sampling process bear no relation to the resources lost in subsequent series production.

In the following practical example, I would like to use a simple calculation to show you how important proper, time-consuming and structured mold sampling can be for your company.

PRACTICAL EXAMPLE:

Given:

- Molded part: Housing cover
- Plastic: PC/ABS
- Monthly quantities: 10.000
- Term: 4 years

Case 1: Sub-optimal sampling to save time	**Case 2**: Optimally executed sampling, including mold and process optimization.
Optimization measures: 0 days	**Optimization measures**: +3 days
Cycle time: 50seconds	**Cycle time**: 40 seconds
Duration of production per month: 138.89 hours ≈ 5.79 days; over 4 years: 277.78 days	**Duration of production per month**: 111.11 hours ≈ 4.63 days; over 4 years: 222.22 days

Differences:

- Per month: 1.16 production days
- Over a total term of 4 years: 55.56 production days

Conclusion:

In case 1, three days of mold sampling were canceled due to time constraints, resulting in a loss of 52.56 days of machine resources.

As you can see, an optimally managed mold sampling process contributes greatly to your machine resources. Based on the practical example shown, I would like to show you a possible way of how you can sustainably carry out mold sampling and can make time savings in mold sampling by adopting a structured and systematic approach.

1.4 Energy Efficiency Starts with Mold Sampling

Every modern injection molding company should aim to reduce its production costs to a minimum and thus continuously increase its production efficiency. In this section, I would like to show you which machine setting errors are very common in practice and how these can have a negative impact on the energy requirements of an injection molding process. I would also like to explain how you can make a major contribution to energy efficiency in the injection molding process by optimizing the machine settings as early as the sampling phase of the injection molding process. The great advantage of more energy-efficient machine settings is that these measures can be implemented easily and, above all, cost-neutrally on the injection molding machine.

In past decades, the requirements imposed in plastics processing companies were limited to economic efficiency and the quality of the molded part. Today, other new requirements, such as energy savings, play a decisive role (Figure 1.3).

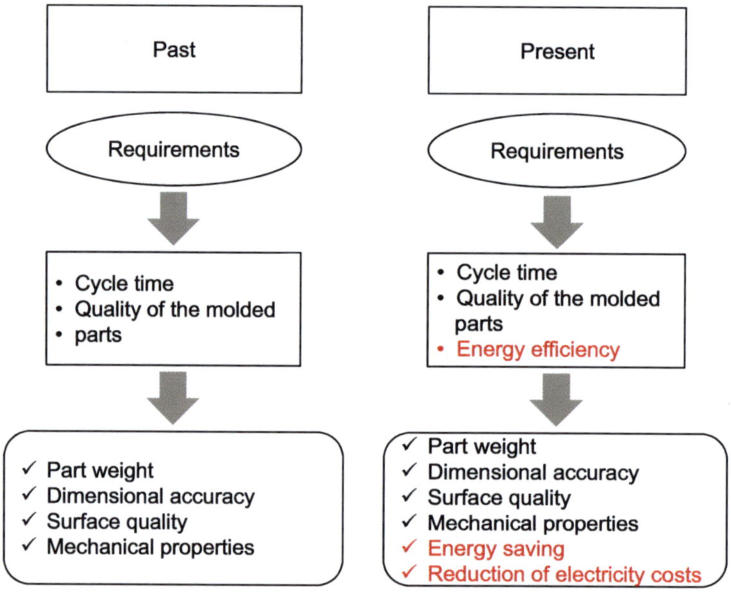

Figure 1.3 New requirements for plastics processing companies

Figure 1.4 shows how electricity prices for industrial electricity (in Germany) have risen sharply in recent years.

In order to remain competitive in the long term, it is important to reduce electricity consumption in the injection molding process while maintaining cycle times and quality in the face of rising electricity prices.

REMEMBER:

Despite all the political changes to the electricity mix, the environmental impact of electricity generation is still very high. Saving electricity is therefore an important step toward reducing the emission of many pollutants that are harmful to the environment and health and toward minimizing the environmental impact.

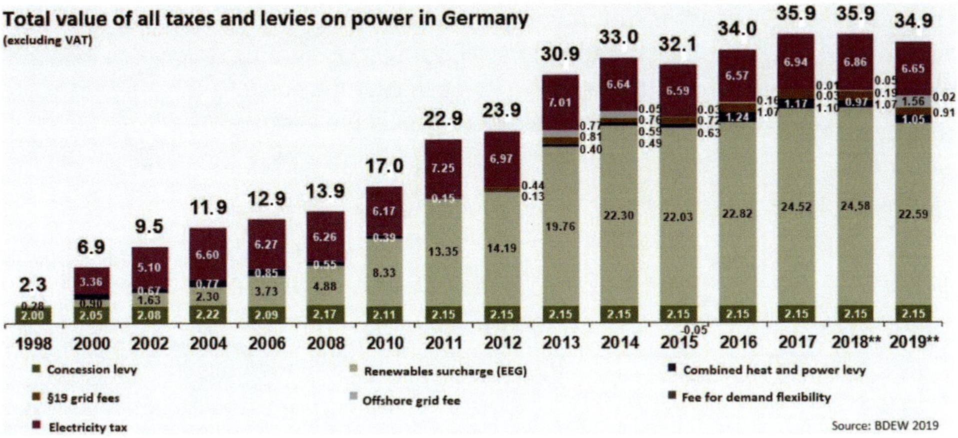

Figure 1.4 Electricity prices for industry (source: BDEW-Bundesverband der Energie- und Wasserwirtschaft e. V.)

1.4.1 Energy and Power Flows in an Injection Molding Machine

To make an injection molding process more energy-efficient, the energy and power flows in the injection molding process must first be considered. Here it is necessary to clarify where the power guzzlers are and then which machine setting parameters can be used to influence power consumption.

Figure 1.5 shows the energy and power flows in a hydraulic injection molding machine (HIM) which are fed into or discharged from the machine during molding or released into the environment via losses.

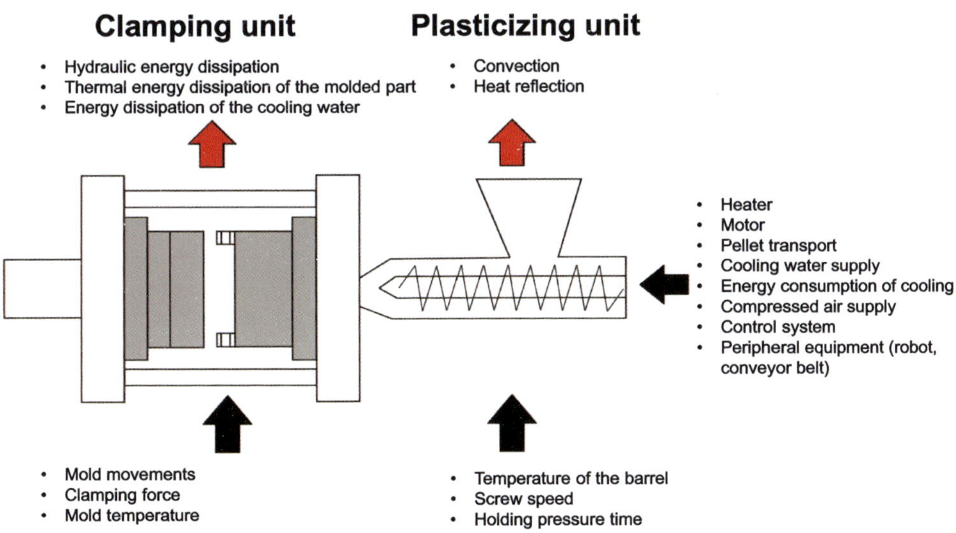

Figure 1.5 Energy and power flows in a hydraulic injection molding machine

The energy and performance influences show that the power guzzlers in injection molding are the clamping unit and the plasticizing unit. To illustrate which machine setting parameters have an influence on this, consider the power consumption for the injection cycle of a hydraulic injection molding machine (Figure 1.6).

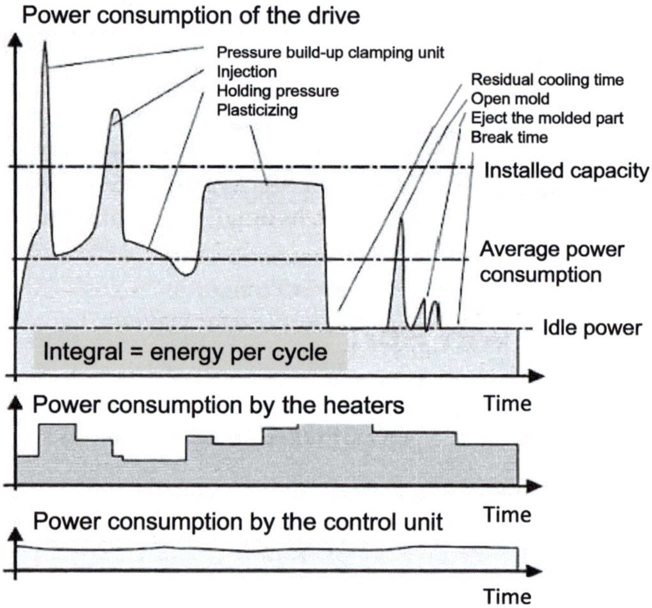

Figure 1.6 Power consumption of a hydraulic injection molding machine over the injection cycle (image source: Stitz S., Keller W.: Spritzgießtechnik, 2nd edition, Carl Hanser Verlag Munich Vienna, 2004, page 330, image number 3.71)

In this example, the power consumption of the drive is plotted against time (injection molding cycle). If you look at the curve, you can see that the plasticizing phase requires the most energy overall. Short-term power peaks also occur during the mold movements and the clamping force (pressure build-up in the clamping unit). In addition to the residual cooling time, holding pressure is another key power consumer in the injection molding cycle. The heating also consumes power during the process and can be reduced by using barrel insulation, for example. The power consumption by the control system is constant but negligible, as it is not a major energy consumer in the cycle.

In the following sections, I would like to explain how, with the right procedure, you can optimize the machine setting parameters in practice in order to save energy.

1.4.2 Energy-Saving Potential of the Plasticizing Unit

1.4.2.1 Barrel Temperature

In practice, injection molding companies very often produce in the upper third of the barrel temperature range recommended by the material manufacturer. There are several reasons for this. On the one hand, mistakes have already been made in the preliminary stages of part and mold design, such as a poorly selected connection concept or the wall thickness/flow ratio has not been taken into account. As a result, right from the start of the molding phase, the “mold sampling specialist” is forced to compensate for the filling problems of the plastic part by setting a higher barrel temperature, for example. The problem can only be solved by a design change to the part or mold. On the other hand, it can be seen in practice that injection molding companies specify fixed material temperatures regardless of the mold, e. g. polypropylene (PP) is always processed at 260 °C. From an energy point of view, the minimum recommended processing temperature of the material manufacturer should be used as a starting point for the molding process. This has the advantage that energy is saved during heating and therefore the molded part has to dissipate less heat. As a result, the required demolding temperature can be reached more quickly and the cycle time can be reduced.

1.4.2.2 Plasticizing (Screw Speed)

The plasticizing phase is one of the most energy-intensive intervals in the injection molding process. However, the screw speed not only has a major influence on the power consumption of an injection molding machine but, in addition to the barrel temperature, also plays a significant role in the quality of the plastic melt to be processed. Unless otherwise specified by the material manufacturer, the screw speed setting is based on the end of the residual cooling time. The speed should only be set so high that the plasticizing process is completed at the end of the residual cooling time.

1.4.2.3 Holding Pressure and Holding Pressure Time

The holding pressure should only be set so high that the molded part does not have any sink marks. In practice, it is very often the case that when the holding pressure time is increased, no further attention is paid to the already set residual cooling time. Consequently, the total cycle time is extended, resulting in higher specific energy consumption. The “mold sampling specialist” should determine the holding pressure time using the sealing point via the plastic part weight. Particular attention must be paid to the fact that, if the holding pressure time is increased, the same increase must be deducted from the residual cooling time in order to keep the injection molding cycle time constant.

1.4.2.4 Residual Cooling Time

The residual cooling time only serves to cool the plastic part in the cavity until a sufficient demolding temperature is reached; this varies one plastic to another. It should be noted that an excessively long residual cooling time has a negative impact not only on the molding cycle but also on the power consumption of the molding machine. The reason for this is that the machine consumes idle power during the residual cooling time. This idle power is the power consumption that is wasted every time the machine stops, e. g. while waiting for the residual cooling time to elapse.

In practice, it is very common to use tables of values or approximation formulas found in the literature to calculate the cooling time. These approximation formulas do not take into account the fact that, when reinforcing materials such as glass fibers are used, the heat is removed from the thermoplastic more quickly, due to the higher thermal conductivity. Accordingly, the residual cooling time of a reinforced thermoplastic can be considerably shorter. As a result, residual cooling times are often set too long in practice. This leads to longer cycle times and thus to higher specific energy consumption. These cooling times determined from value tables and approximation formulas may only be used as initial values for mold sampling and do not constitute fixed values for later production. In this case, the "mold sampling specialist" should start with a longer residual cooling time and reduce it continuously over the course of several injection molding cycles. It is important that the demolding temperature of the molded part be measured after each shot. An optimum residual cooling time is set when all areas of the plastic part have reached the demolding temperature recommended by the material manufacturer.

1.4.3 Energy-Saving Potential of the Clamping Unit

1.4.3.1 Clamping Force

An important parameter is the clamping force of the injection mold. One problem here is that the maximum possible clamping force of the injection molding machine is very often used in production. A clamping force that is set too high not only consumes a great deal of energy, but also promotes mold wear due to high surface pressure. Furthermore, venting problems can occur on the mold if the clamping force is set too high. The "mold sampling specialist" should therefore always determine the required clamping force.

1.4.3.2 Mold Movements

If we consider the mold movements alongside the excessive clamping forces, the opening and closing movements are usually not taken into account as a possible source of energy savings. The fast movements of the clamping unit with mold must be accelerated and decelerated again under a very high energy input. Consequently, the mold movements should be individually adapted to the respective mold. It is important here to set the optimum opening stroke required. This saves cycle time and, as a result, energy, due to shorter travel distances of the clamping unit.

1.4.3.3 Mold Temperature

In injection molding plants, it is often the case that injection molds are set to the maximum temperatures recommended by the material manufacturer. This phenomenon is subject to the same problems as those already discussed in the section on barrel temperature. Here too, design errors may have been made in the preliminary stages of mold and part design. As a result, the "mold sampling specialist" must also select a higher mold temperature in order to compensate for filling problems, for example. In practice, it is very common for many injection molding companies to specify a certain temperature value, e. g. the mold temperature for polypropylene (PP) is always set to 60°C. Here, the minimum recommended mold temperature should be used first. In this case, you should start with the minimum mold temperature recommended by the material manufacturer. This has the great advantage that energy can be saved during heating and at the same time the required average demolding temperature of the plastic part can be reached more quickly.

1.4.4 Specific Power Consumption

The specific power consumption provides a reliable way of comparing machines, settings and processes. The specific power consumption of an injection molding machine is defined as the power consumption per kilogram of material (kWh/kg).

Theoretically, a specific power consumption of 0.1 kWh/kg can be achieved, but such values cannot be realized in practice and are 2–3 times higher. This is due not only to the technology of the injection molding machine, but also to insufficient machine utilization, e. g. shot weights that are too small in relation to the machine's capabilities. Today, under favorable conditions, a specific consumption of 0.5 kWh/kg can be achieved with fully hydraulic drives and of 0.2 kWh/kg with fully electric drives. Chapter 8 "Optimizing the basic setting" goes into more detail on an assessment and possible optimization options.

1.4.5 Conclusion on Energy Efficiency During Sampling

Industrial electricity prices in Germany have almost doubled in recent years and will continue to rise. As a result, power consumption now makes a decisive contribution to the manufacturing costs of a molded part in the company.

As you can see, energy optimization can make a major contribution to more energy-efficient production, simply by taking certain machine setting parameters into account during mold sampling. Depending on the operational situation, it is possible to save between 10% and 30% energy simply by optimizing the machine setting parameters. This sensible use of energy resources plays a decisive role in your economic fortunes.

The aim of every mold sampling process should be to achieve consistent molded part quality and the same cycle time with significantly lower power consumption.

REMEMBER:

The higher the set machine process parameter, the greater is the power consumption.

1.5 Using Injection Molding Simulations Effectively for Sampling

Simulation technology is still under-used for mold sampling

Today, injection molding simulation has developed into a standard tool. The entire injection molding process can be simulated from filling to holding pressure and holding pressure time through to warpage, orientations, injection pressure, injection time, cycle time and temperatures (see Figure 1.7).

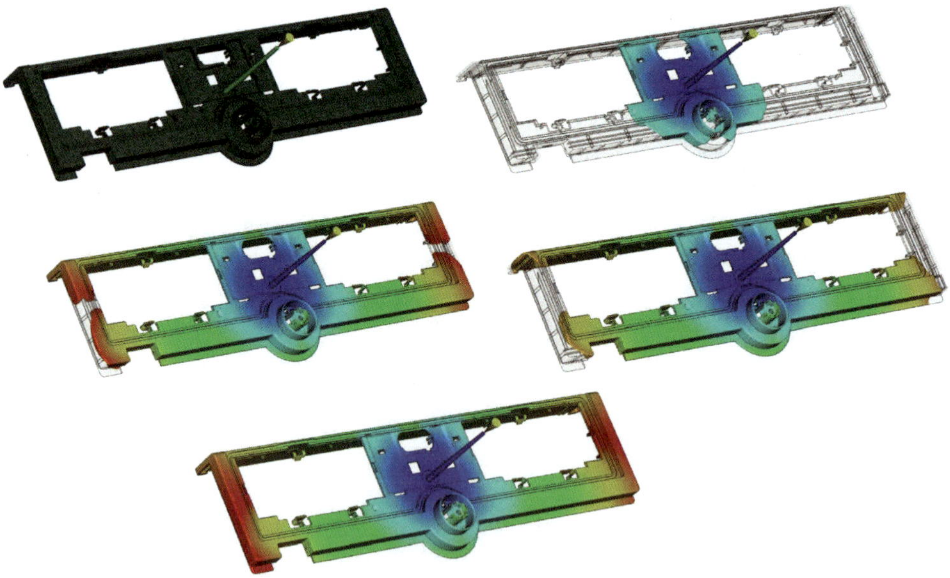

Figure 1.7 Injection molding simulation with FEM programs

In practice, the recommended process parameters are taken from the material data sheets supplied by the raw material manufacturers, and that is absolutely the correct thing to do. However, the recommended values are usually specified within a certain range, such as a specific holding pressure of 300–550 bar.

Therefore, a future-proofing approach—if a simulation of the part has been performed—would be to use the process parameters obtained from a simulation as possible guide values or reference points for the upcoming mold sampling.

The biggest advantage here is that the process parameters obtained from the simulation will have been calculated individually for the molded part to be sampled. Consequently, important criteria for the selection of process parameters are given in advance, taking into account geometry, wall thicknesses, molecule and fiber orientations, mold and melt temperature. Another advantage is that possible difficulties with the part can be recognized at an early stage and passed on to the "mold sampling specialist".

In the coming years, the trend will be for customers of molded parts to increasingly demand simulations from suppliers, and this will also have a positive effect on mold sampling.

2 Information Gathering and Preparation for Mold Sampling

This chapter deals with the gathering of information and the correct preparation for mold sampling. With the help of these two points, you can save a lot of time and resources in the run-up to the sampling process, as, for example, unnecessary walking or long searches for data can be avoided.

2.1 Information Gathering

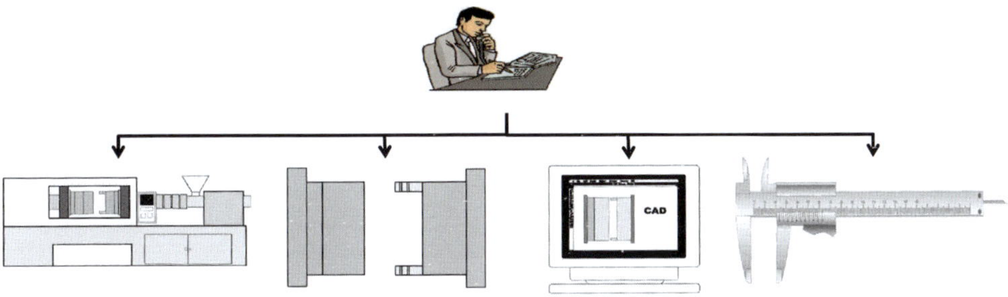

Figure 2.1 General information gathering

General information gathering is the first step. Here, the following information is clarified by the "project manager" and passed on to all employees involved in sampling.

- What is to be sampled (article designation, mold number)?
- When should this take place?
- What type of mold sampling is to be implemented (e. g. initial, subsequent, material sampling)?

- What is the current molded part index?
- Is an approved and current molded part drawing available?
- Are there already mold sampling documents?

The second step is to obtain information about the machine and peripherals, the material and the number of tooling samples required and their packaging. The following points must be checked by the "project manager" and passed on to the "mold sampling specialist":

- Which injection molding machine is to be used for sampling?
- Is the machine suitable for the injection mold to be sampled (clamping force, installation height, max. ejector stroke, max. plasticizing stroke, etc.)?
- Is there a suitable way to remove the molded parts from the machine (handling system)?
- Has all necessary maintenance work been carried out on the machine? (machine capability (MFU) and calibration according to VDMA 24.470, Parts 1 + 2)
- Has the machine been running trouble-free recently?
- What kind of material is needed?
- Does the material need to be colored? If so, with which masterbatch?
- Is the material manufacturer's recommended processing data available? If not, must this data be requested from the material manufacturer?
- Does the material need to be pre-dried? If so, how long and at what temperature?
- How many tooling samples are to be produced?
- How should the tooling samples be packaged?
- Are there any special features of the mold or the sample part that need to be taken into account?
- Is a mold cooling plan available?
- Are results from an injection molding simulation available?

The questions listed above are a necessary part of the initial process and ensure that all required information is collected and passed on to the employees involved.

2.1.1 Information Sheet for Sampling

Unfortunately, current practice shows that this information is usually only passed on verbally over the phone or by writing in a short email. As a result, the first communication errors occur in the early phase of mold sampling, and this can then lead to considerable problems in the further sampling process.

Information sheet for mold sample

General information from the project manager

Article description:		**Type of sampling:** ❒ Initial sampling (FOT) ❒ Follow-up sampling ❒ Material sampling ❒ Machine change ❒ Process optimization
Item number:		
Mold number:		
Customer:		
Sampling date:		
Machine number:		

Information for machine operator

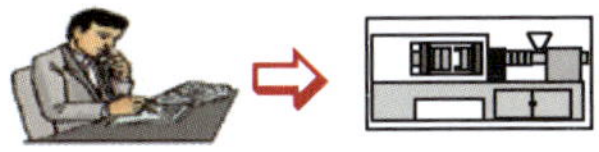

Injection molding machine:		
Series/type:		
Machine suitability (max. installation height, max. clamping force, max. ejector stroke, max. plasticizing stroke, etc.) checked?		❒ Yes ❒ No
Run and rate data:		
Calculated cycle time:		
Calculated shot weight:		
Fixed quantities:		
Sample parts:		
Number of samples required:	part	Type of packaging:
Samples for the attention of quality	part	
Instructions/important things to note:		
Comment:		

Information for quality assurance/measurement technology specialist

Current drawing status available:	❒ Yes ❒ No
Drawing status/index:	
Comment:	

Information for moldmaker/toolshop

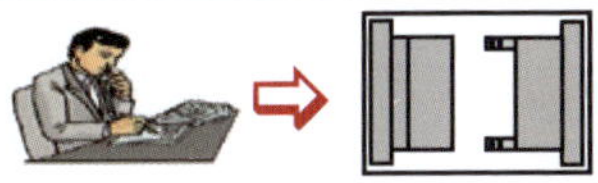

Comment:

Information for the design engineer

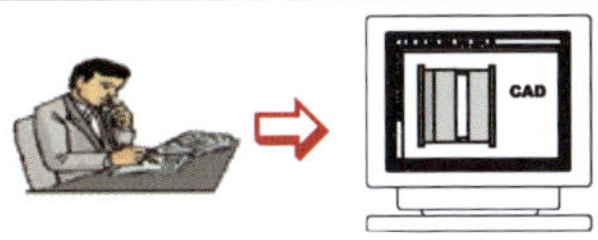

Comment:

Figure 2.2 Information sheet for sampling

In order to guarantee a constant flow of communication within the company, it is advisable to create an "information sheet for sampling" that summarizes all the important information. This document is completed by the "project manager" before each sampling and forwarded to all employees involved in sampling within the company. The "information sheet for sampling" has the advantage that a separate section is labeled with the corresponding symbol for each person involved. This means that the "project manager" can pass on the necessary information specifically for each individual employee in one document.

The layout of an "information sheet for sampling" is illustrated in Figure 2.2 with an example.

2.2 Preparation for Mold Sampling

Once the "mold sampling specialist" has received the information sheet on pending sampling, the "mold sampling checklist" (Figure 2.3) can be used for sampling preparation at the injection molding machine.

Mold sampling checklist

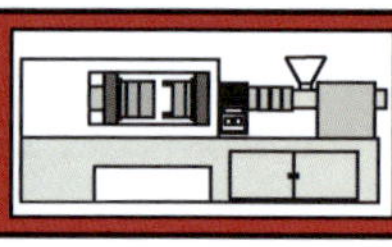

Description of the article:		
Item number:		**Type of sampling:**
Mold number:		❒ Initial sampling (FOT)
Customer:		❒ Follow-up sampling
Date of sampling:		❒ Material sampling
Machine number:		❒ Complaint
Machine operator:		❒ other

Please document all deviations (fields ticked with "No") in the sampling report with reasons/comments.

Gathering information and preparing the sampling				
1	Information sheet for sampling available and completed in full?			**Block 1**
2	Preparation for sampling on the injection molding machine completed?			

Figure 2.3 Mold sampling checklist: block 1

Mold sampling includes providing all documents and data as well as the necessary accessories at the injection molding machine (see Figure 2.4). The following documents are needed at the injection molding machine:

Table 2.1 Overview of Mold Sampling Documents

Information sheet for mold sampling	Section 2.1.1, Figure 2.2
Mold sampling checklist	Information on the book layout
Technical material data sheet and processing instructions from the material manufacturer	See example: Section 4.4.1 "Setting the mold temperature", Figure 4.11
Mold cooling plan	See example: Section 3.2.1 "Mold installation procedure", Figure 3.5
Water connection plan	Section 3.2.1 "Mold installation procedure", Figure 3.6
Determination of the required holding pressure time	Section 6.2.2 "Holding pressure time", Figure 6.6
Experimental optimization of the mold clamping force	Section 6.3.1 "Experimental optimization of the clamping force", Figure 6.12
One-factor method/design of experiments (DoE)	Section 8.1.2.1 "Carrying out molding tests by the one-factor method", Figure 8.5, Figure 8.7
Evaluation of the injection molding tests	Section 8.1.3.1 "Template for evaluating molding tests", Figure 8.9
Optimization of energy efficiency	Section 8.2.3 "Optimization options for increasing energy efficiency", Figure 8.19
Process capability analysis of machine setting and process parameters	Section 8.3.1 "Step 1: Process capability analysis of machine and process parameters", Figure 8.21
Process Run@Rate of the optimized basic setting	Section 8.3.2 "Step 2: Process Run@Rate of the optimized basic setting", Figure 8.24
Machine setting data sheet	Section 9.2 "Documentation of the machine setting and process parameters", Figure 9.2
Mold sampling report	Section 9.3 "Mold sampling report", Figure 9.3
Mold hand-over protocol	Section 11.3.1 "Final meeting for mold hand-over to series production", Figure 11.3

You should also have the following tools and equipment ready to hand at the injection molding machine:

- The injection mold to be sampled
- The plastic to be processed and, if necessary, the corresponding coloring agent (masterbatch)

- Protective gloves and safety goggles
- The corresponding drawing for the molded part
- Measuring devices:
  - A dial gauge to determine the required mold clamping force to an accuracy of 1/1000 mm
  - A scale that works to at least three decimal places and an accuracy of 0.001 g
  - A temperature-measuring device and a thermal camera (thermography)
  - A caliper gauge
  - A spirit level
  - Radius gauges and a plug gauge
  - A water flow meter
- Water hoses
- Your laptop or tablet
- A pocket calculator

Figure 2.4 Preparing for sampling at the injection molding machine

Milestone: "Information gathering and preparation for sampling"

This has been reached once you have completed block 1 of the "Mold sampling checklist".

3 Mold Set-Up

3.1 Before Installing the Mold

For this purpose, the "mold sampling specialist" must be entered as the first working step in the mold sampling checklist, Block 2 ("Mold set-up").

Mold set-up				
Before fitting the mold in the machine:		**Yes**	**No**	
3	Safety checks carried out on the injection molding machine (crane or peripherals) and no defects found?			
4	**Checking the injection mold:**			
4.1	Mold transport bridge present and mounted?			
4.2	Are the labels on the injection molds present and visible?			
4.3	Are there devices for placing the injection molding tool on the floor (optional)?			
4.4	Mold free of visual damage?			
4.5	Is the jaw or slider movement (locking) limited?			
4.6	Ejector plate safety device available?			**Block 2**
4.7	Have the mold installation dimensions (mold centring Ø nozzle side), installation height, length of the ejector ram, sufficient machine tie bar distance, etc.) been checked?			
4.8	Cooling circuits continuous (compressed air test)?			
4.9	Are the mold temperature control connections for the inputs and outputs correctly labelled?			
4.10	Flow rate (l/min) checked and documented in the machine setting log?			
For hot runner molds:				
4.11	Functional test of the hot runner system (HKS) performed and OK?			

Figure 3.1 Mold sampling checklist: Block 2

3.1.1 General Safety Checks

First of all, it is very important to find a safe workplace for the sampling. To do this, the "mold sampling specialist" reviews the following steps and questions.

- **Step 1:**

  Visual inspection, for example, for water and hydraulic leaks, improper or damaged wiring of the injection molding machine and peripherals, defective machine components, porous or damaged hydraulic hoses, etc.

- **Step 2:**

  Functional check of all safety elements of the machine and peripherals, such as safety doors and emergency stop switches.

- **Step 3:**

  Checking the lifting crane:

  - What is the maximum load capacity of the crane system and has it been tested by an independent test institute?
  - What is the maximum load capacity of the crane chains and hooks and have they been tested by an independent test institute?
  - Functional check of emergency and main switches
  - Functional check of all travel movements in x, y and z directions
  - Visual inspection for damage to wire ropes, such as kinks, broken strands, loosening of the wire mesh, crushing of the rope, corrosion scars, signs of wear, etc.
  - Visual inspection for damage to steel link chains, such as worn or deformed chain links
  - Visual inspection for damage to the crane hook, such as cracks or transverse cracks in the neck, shank, thread or hook mouth
  - Functional check of the fail-safe device of the crane hook

- **Step 4:**

  Checking the required eyebolts:

  - Is there any visible damage?
  - Compare the total weight of the injection mold with the maximum permissible load capacity of the crane
  - Is the eyebolt screwed in to the correct depth?

NOTE:

The correct screw-in depth should be approx. 1.5 × Ø of the metric ISO thread of the eyebolt.

3.1.2 Checking the Injection Mold

Following the safety checks, the "mold sampling specialist" checks the injection mold with regard to the following issues:

- Is the tool transport bridge present and correctly installed?
- Are all mold labels present and clearly visible?
- Are the support feet for the mold present (optional)?
- Is the mold free of visible damage?
- Is there a locking or safety device for the splits or the slider in the mold?
- Is the ejector plate safety device in place?
- Checking important mold dimensions:
  - Comparison of the minimum installation height of the injection molding machine with the length of the mold
  - Comparison of the mold width with the tie bar spacing of the injection molding machine
  - Comparison of the diameter of the mold centering ring (nozzle side) with the mold centering on the machine clamping plate (nozzle side)
  - Comparison of the sprue bushing radius (mold) with the nozzle radius (machine) using a radius gauge

> **NOTE:**
> The radius of the machine nozzle should always be smaller than the radius of the sprue bushing of the mold.

  - Measure the diameter of the runner (mold) with a caliper gauge or plug gauge → compare with the measurement of the nozzle bore diameter (machine)

> **NOTE:**
> The bore diameter of the machine nozzle should always be 0.5 mm–1 mm smaller than the bore diameter of the sprue bushing of the mold.

  - Measure and check the length of the required ejector ram on the machine
- Are all mold cooling circuits continuous?

NOTE:

Checking the continuity of the cooling channels with compressed air "before installation" is strategically very important. In the event of non-continuity, the mold can be returned to the mold-maker/toolshop for correction without unnecessary mold set-up work, saving valuable time! In the case of new molds, it can happen that metal chips are still present or, in the case of molds that have already been put into operation, limescale deposits come loose. This is why it is important that the "mold sampling specialist" wears protective goggles and holds a cloth in front of the cooling bore to be inspected. This ensures that no particles are blown outwards uncontrollably.

- Are the circuit designations correctly labeled with the on and off points of the cooling channels?

3.1.3 Sampling of Hot Runner Molds

3.1.3.1 General Information on the Sampling of Hot Runner Molds

A lack of knowledge regarding the correct start-up of hot runner systems can lead to damage to the heating cartridges, for example. To get these problems under control, a functional test and a detailed procedure for the correct start-up of empty hot runners are shown below.

BACKGROUND KNOWLEDGE:

Ceramic cartridge heaters are frequently used in the plastics processing industry. If the injection mold with integrated hot runner system has not been used for a long time, moisture may have accumulated in the heating cartridges. If full power is demanded of the heating cartridges during heating, high temperatures are generated very quickly. These high temperatures lead to an almost explosive vaporisation of moisture which can damage the cartridges. This should be taken into account when commissioning the hot runner system for the first time, just as with an existing series mold.

3.1.3.2 Initial Commissioning and Functional Check of the Hot Runner System

A functional test of the hot runner system "before installation" of the mold is very important. If the hot runner connectors are damaged or incorrectly connected, the mold can be returned to the mold shop for correction without unnecessary mold set-up work, thus saving valuable time.

In order to guarantee gentle heating of the hot runner system, the "mold sampling specialist" must use a hot runner heating ramp (Figure 3.2). Here it is important not to immediately set the target value recommended by the material manufacturer (e.g. 240 °C), but to start gradually with a lower holding value (e.g. 100 °C). The holding value is maintained for a defined time of 10–15 minutes. During this time, the moisture can slowly escape from the heating cartridges and damage can be prevented. It must also be checked that all the set heating zones of the hot runner system heat up smoothly and evenly.

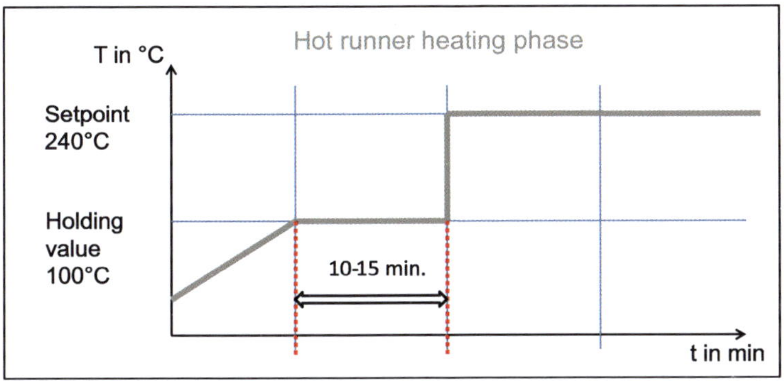

Figure 3.2 Example of a hot runner heating phase

> **NOTE:**
> All newer makes of hot runner control units usually have a type of integrated "assistance function" or automatic controlled heating function (soft start and heating ramp).

3.1.3.3 Procedure for Starting Up and Filling an Empty Hot Runner System

In this section, I would like to show you step by step how the empty hot runner should be filled and what to look out for before the first mold sampling.

- **Step 1:**

  The injection mold and the barrel must have reached the recommended temperatures

- **Step 2:**

  Switch on the hot runner controller and start the hot runner system with a heating ramp, as shown in Figure 3.2 at, e.g. 100 °C, for 10–15 min. Many newer hot runner controllers have an integrated start-up circuit; here, ramped heating is carried out from the controller

- **Step 3:**

  Set the processing temperature recommended by the material manufacturer on the hot runner controller and allow the hot runner system to warm up for approx. 5 minutes

- **Step 4:**

  Check that the valve gate is functional

- **Step 5:**

  Advance the plasticizing unit until the barrel nozzle is in contact with the sprue bushing of the injection mold. It is important that the barrel nozzle be correctly zeroed and the contact pressure for the plasticizing unit built up. Otherwise there may be leaks between the barrel nozzle and the mold sprue bushing

- **Step 6:**

  Fill the hot runner system with the barrel nozzle in contact and the shut-off nozzles open at a limited specific injection pressure (e. g. 350–400 bar) until the material emerges at the gates in the mold. It is important that natural material (without coloring agents or glass fibers) be used for the first filling of the hot runner system

3.2 Mold Installation

The "mold sampling specialist" continues to work on block 3 in the mold sampling checklist (Figure 3.3) as the next working step. The specialist must check the following important points here.

Mold fitting:		Yes	No	
5	Eyebolt in the mold center of gravity?			Block 3
6	Mold mounting grooves with sufficient clamping surface?			
7	Correct screw-in depth (approx. 1.5 × Ø) of the screws observed?			
8	Can mold cooling be connected without any problems?			
9	Can the nozzle side and ejector side be removed ?			

Figure 3.3 Mold sampling checklist: Block 3 "Mold installation"

3.2.1 Procedure for Mold Installation

Optimum mold installation should proceed as smoothly as possible. The following description illustrates optimum mold installation.

- **Step 1:**
  - In set-up mode, move the clamping unit to the maximum opening stroke with "Open mold"
  - Lift the injection mold with the lifting crane
- **Step 2:**
  - Check the cooling circuits for continuity using compressed air

> **NOTE:**
> If the mold is too large, it must first be fully installed in the injection molding machine. Then, the cooling system on the underside of the mold must be checked for continuity. For your own safety, you should never step under a suspended mold.

  - **Only for molds with a hot runner system:** Switch on the hot runner heating and check that the hot runner system is functioning properly (allow for the hot runner start-up phase)
- **Step 3:**
  Lift the injection mold into the injection molding machine and watch out for possible collisions with the machine's tie bars
- **Step 4:**
  - Center the mold on the nozzle-side clamping plate of the injection molding machine and tighten it loosely with a fastening screw or clamp (fastening element). To ensure that the mold is optimally aligned, check the mold with a spirit level or correct the alignment
  - Fit the correct ejector rod to the mold
- **Step 5:**
  - Slowly close the clamping unit in set-up mode with "Close mold"
  - Zero the mold in the machine when it is closed (see Figure 3.4)
  - Take some of the load off the lifting crane
  - Now fully attach or clamp (fasten) the injection mold on the operating side and on the rear side. Ensure that the screws are screwed in to the correct depth
- **Step 6:**
  Release the crane hook from the injection mold and move the lifting crane out of the machine area

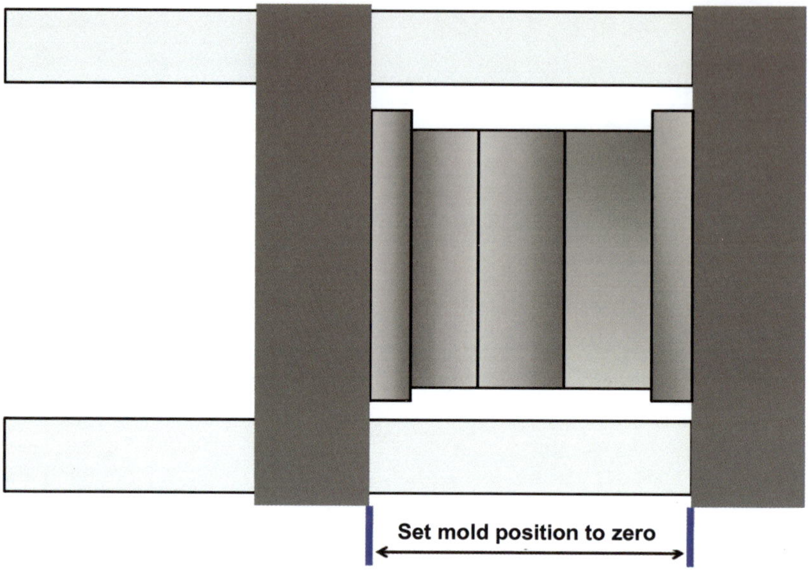

Figure 3.4 Zeroing the mold

- **Step 7:**

  All connecting elements of the injection mold, such as the mold transport bridge or a mold opening safety device (usually marked in red), must be removed

- **Step 8:**

  Slowly move the injection mold to the maximum opening stroke in set-up mode

> **NOTE:**
>
> When the mold is opened for the first time, it must be checked for ease of movement. This is particularly important for functional parts, such as splits or slides.

- **Step 9:**

  Connect the mold temperature control flow and return as well as the hose bridges to the mold. For this purpose, the mold cooling plan (Figure 3.5) should be available at the machine to provide information for connecting the mold temperature control.

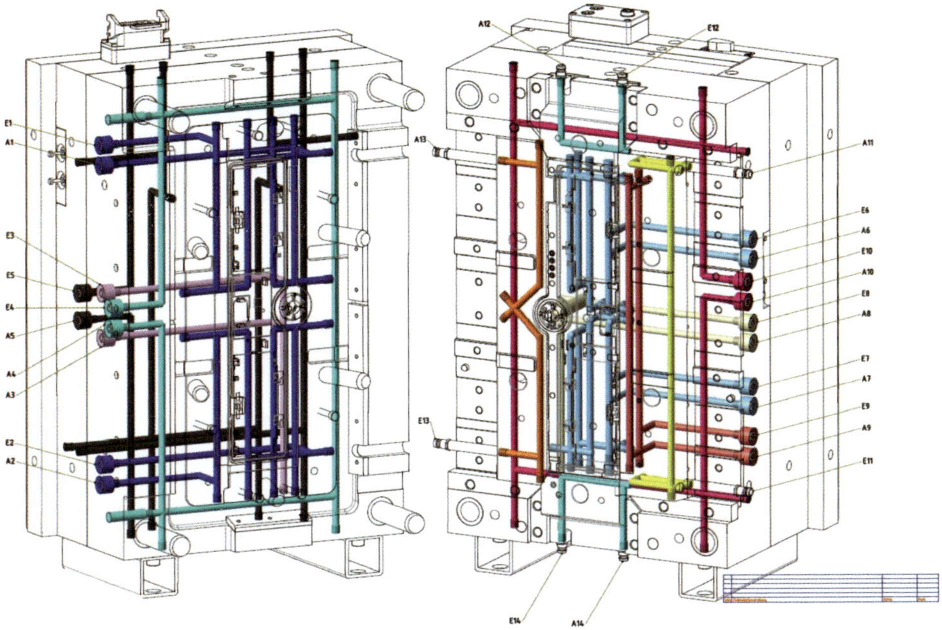

Figure 3.5 Example of a mold cooling plan (image source: Gentner Kunststoffverarbeitung GmbH)

NOTE:

The following details must be observed to ensure that the mold temperature control hoses are connected correctly:

- Separate temperature control of the nozzle side and the ejector side
- In order to be able to run different flow temperatures, separate circuits should be connected for sliders/splits and cores
- For greater process reliability, the temperature control circuits should only be connected in series and not in parallel
- For higher mold temperatures (> 60°C), the hoses should have hose insulation to increase energy efficiency
- Care must be taken to ensure that the hoses are connected without twists, bends or kinks
- The cooling water supply and return hoses must not follow any unnecessary paths from the machine to the temperature control unit.
- To avoid unnecessary pressure losses in the mold cooling channel system, the water hoses must not be too long. The required length of the hoses should be individually adapted to the set mold movements
- The freely suspended water bridges must be arranged in such a way that the bridges do not come into contact with machine parts when the mold is running

To ensure that the water circuits are always connected in the same way during subsequent series production with the mold, all connected water circuits on the nozzle and ejector side must be documented. A "water connection plan" template (Figure 3.6) has been designed for this purpose. The "mold sampling specialist" can use this to document all connected circuits, such as

- Flow—El—A1
- A1—E3
- A3—E4 etc.
- Connect the ejector plate safety cable (if required) to the mold

- **Step 10**:
  - Couple the ejector rod with the machine's ejector coupling
  - Zero the ejector in the rear position (retracted) on the machine

Milestone: "Mold set-up"

- You have worked through block 2 of the mold sampling checklist.
- The mold set-up process is complete when you have correctly carried out and checked all the previous steps.

Water connection plan		
Article description:		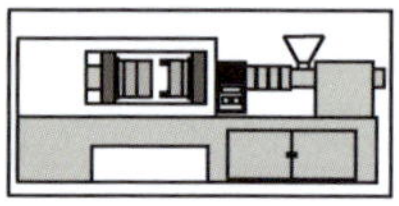
Mold number:		
Date:		
Created by:		

Nozzle side or fixed side:

Ejector side or moving side:

Figure 3.6 The "water connection plan" template

4 Basic Setting of the Clamping Unit

This chapter deals with the clamping unit. The starting assumption here is that the mold on the machine is fully set up and connected with the required accessories, such as temperature control medium, hot runner, ejector plate safety device, pneumatics or hydraulics. For the adjustment of the clamping unit, the "mold sampling specialist" works through block 4 ("Basic adjustment of the clamping unit") in the "Mold sampling checklist", see Figure 4.1.

Basic setting of the clamping unit				
Mold movements:		**Yes**	**No**	
10	Moving mold parts lubricated?			**Block 4**
11	Moving mold parts (slider, ejector, etc.) running smoothly?			
12	Mold movements (travel paths, speeds, ejectors, additional functions in the mold (cores etc.)) set?			
12.1	Mold zeroed?			
13	Mold and checked for function?			
14	Basic setting of the mold clamping force calculated and set?			
15	Correct mold temperature set?			

Figure 4.1 Mold sampling checklist: Block 4

4.1 Mold Movement

4.1.1 Opening the Mold

The mold opening stroke *x* (Figure 4.2) should be selected so that parts can be demolded and then removed without any problems. It is possible to achieve an economical injection molding cycle by avoiding unnecessarily long opening strokes by the injection mold.

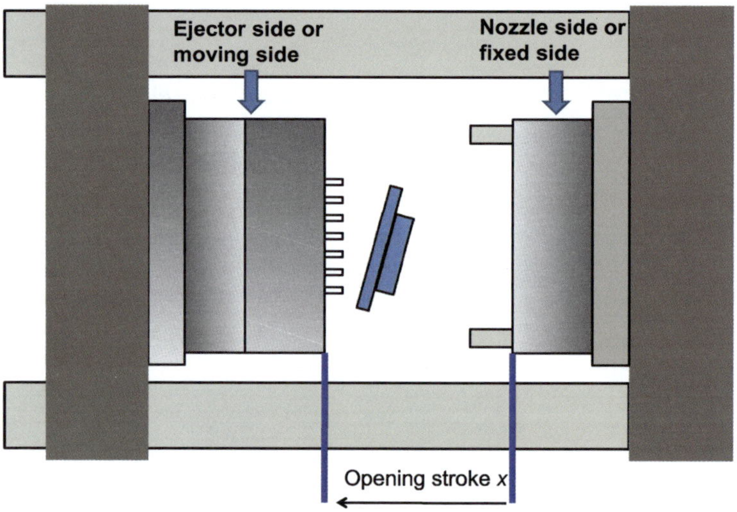

Figure 4.2 Mold opening stroke *x* (schematic)

4.1.2 Mold Opening Speed

In order to ensure safe, wear-free and mold-friendly opening in the injection molding process, a three-stage speed profile should be used that is individually adapted to the injection mold. In the first stage, it is important to ensure that a slow movement is initially selected. This is the case when the molded part has been completely demolded from the nozzle side of the mold or when the guide pillars of the nozzle side no longer protrude into the ejector side of the mold in the case of split or slide molds. In the second stage, a fast movement is set until shortly before the set mold opening stroke (Figure 4.2). A further, third stage is used for gentle braking of the mold. The aim should be smooth, jerk-free mold movement. Figure 4.3 shows an example of a three-stage speed profile.

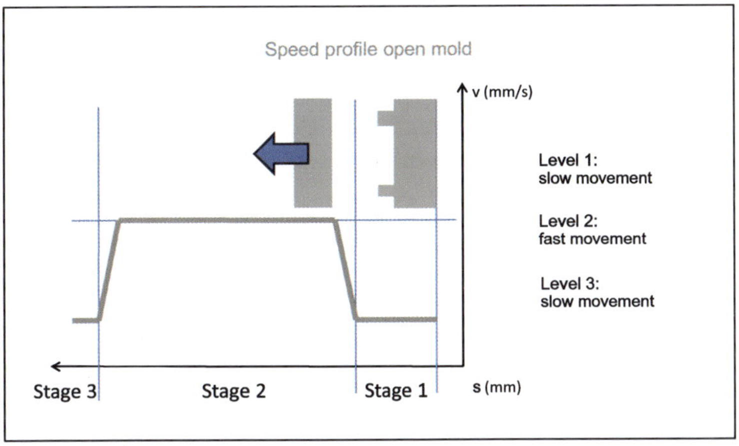

Figure 4.3 Speed profile for "Open mold"

4.1.3 Closing the Mold

The closing movement must also be individually adapted to the injection mold. In order to guarantee reliable closing in the subsequent process, a speed profile must also be used here. The profile begins with a short, slow first stage. The second stage follows with a fast movement until shortly before the ejector side makes initial contact with the mold half on the nozzle side. Depending on the mold design, this contact is the immersion of the guide pillars or, in the case of split/slide molds, in the inclined guide pillars. From this point onwards, a slow, third stage must be used until the mold is closed. Here too, the speed profile should be set such that smooth, jerk-free mold movement is ensured. Figure 4.4 shows a three-stage speed profile for "Close mold".

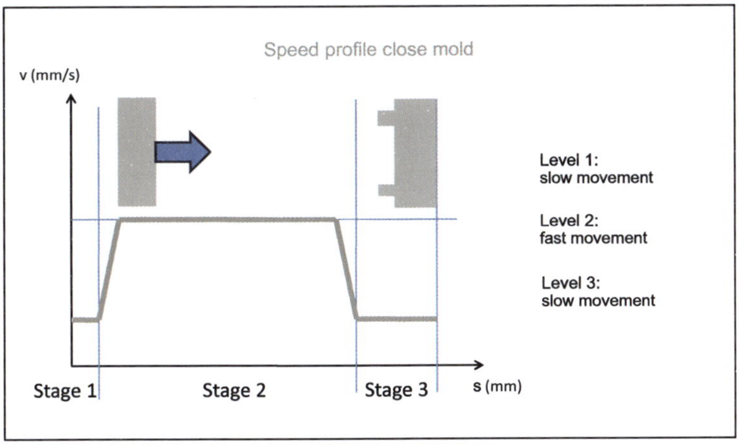

Figure 4.4 Speed profile for "Close mold"

4.1.4 Setting the Mold Ejectors

The ejector stroke and ejector speed must be adapted to the plastic part. Care must be taken here to avoid an unnecessarily long ejector stroke and to ensure that a suitable advance and retraction speed contributes to mold-friendly and wear-reduced injection molding.

4.1.5 Additional Functions in the Mold

All existing additional functions in the mold, such as the hydraulic or electric core pulls, must be checked for functionality. Here it is important to create a "sequence program", e. g. for the opening and closing times of the individual core pulls in the machine control system. This sequence program contains all the necessary movements of the core pull for the entire cycle. If an incorrect sequence program is set, this can cause major damage to the mold.

4.2 Mold Protection

In practice, optimally set mold protection is often only given marginal consideration. Although speed profiles are taken into account when closing the mold, the clamping force and monitoring time are usually not set optimally or are neglected.

BACKGROUND KNOWLEDGE:

Mold protection prevents the mold from being damaged during closing if a foreign object or a non-demolded part gets stuck in the parting line or cavity area. This safety device also protects against damage, e. g. in the case of slide or split molds. Here it can happen that a slide (or split) does not remain in its home position when the mold is closed and the mold causes considerable damage to the slide (or split) and the inclined guide pillars when it is closed. The mold protection works by setting a pressure limit in the clamping cylinder of the injection molding machine. The clamping force is reduced after a set closing stroke, such that the machine opens the mold immediately or stops and consequently signals a fault if there is resistance (e. g. a part that has not been demolded) beyond a set monitoring time.

The starting point of the mold protection is adapted to each injection mold. An important criterion here is that the set clamping force must be effective from the closing stroke at which the moving side first comes into contact with the guide pillars of the fixed side, for example. Figure 4.5 shows a schematic representation of the required effective range of the mold protection.

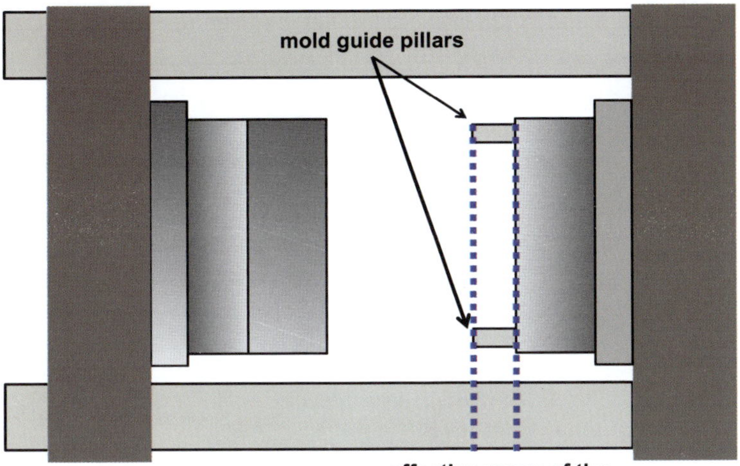

Figure 4.5 Required effective range of the mold protection

4.2.1 Setting Up Mold Protection

All the important procedures for setting up mold protection are explained here step by step:

Step 1: Determining the mold protection stroke

You must first define the starting point of the mold protection stroke (Figure 4.6).

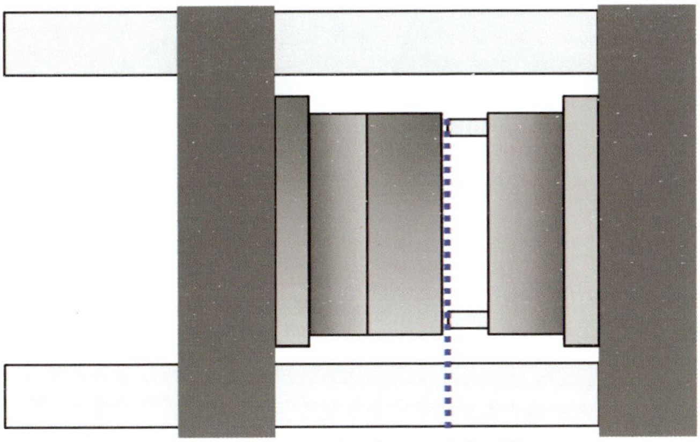

Figure 4.6 Starting point of the mold protection stroke

This starting point is also the beginning of the last level 3 of the speed profile for "Close mold" (see Figure 4.4).

The end of the mold protection stroke (Figure 4.7) must be set to 0.5–1 mm before the mold halves make contact, so that the injection molding machine has time to build up the required clamping force.

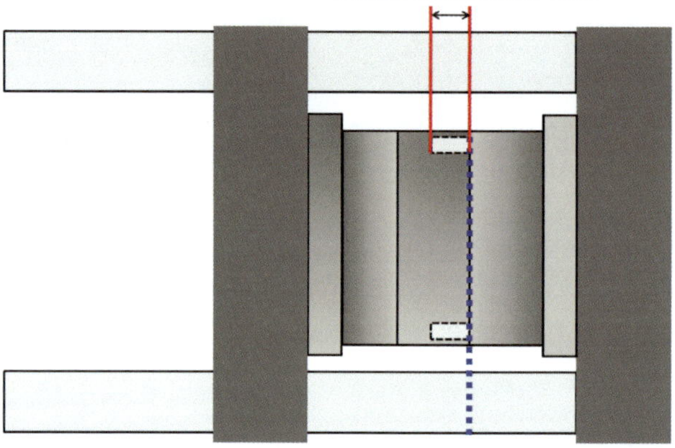

Figure 4.7 End of the mold protection stroke

Step 2: Determining the mold protection stroke

From the start of the defined protection stroke, the minimum clamping force at which the injection mold can still be closed must be checked. This determined minimum force serves as our clamping force, which we set on the machine. If, for example, there is a foreign object in the mold later in the process, the set minimum clamping force can no longer close the mold and will thus protect it from damage.

Step 3: Determining the mold protection time

The mold protection time results from the length of the mold clamping stroke and the closing speed. If there is an obstacle or foreign object in the mold, for example, the machine must apply a higher clamping force in order to close the mold. Due to the finely adjusted mold clamping force, the obstacle or foreign object cannot be overcome and so the mold stops.

Furthermore, a defined time is set by way of a monitoring function. If this time is exceeded, the machine stops and signals a fault.

4.2.2 Checking that Mold Protection is Functioning

The following procedure is recommended for checking that mold protection is functioning:

- **Step 1:**

  To check that the set mold protection is functioning, insert a sprue (or piece of cardboard or folded paper) into the parting line of the mold (Figure 4.8)

Figure 4.8 Performing a functional test of mold protection using a sprue

- **Step 2:**

  Close the mold under high machine pressure (no set-up mode) (Figure 4.9)

- **Step 3:**

  The mold protection is set correctly if the mold clamping force is not great enough to close the mold within a defined mold closing time when “Closing the mold” with an inserted foreign object (e.g. sprue, cardboard or folded paper). The machine must signal a machine fault and open the mold. If this is not the case, the force and duration of mold protection must be optimized again

Figure 4.9 Closing the mold under high machine pressure (no set-up mode)

> **NOTE:**
>
> When checking the function of the mold protection, make sure that no foreign bodies (e.g. sprue, cardboard) get into the cavity area, especially in the case of polished cavity surfaces. The risk of damage in the form of scratches or dents in the cavity surface is very high. To ensure a safe functional test, it is advisable to place the foreign body around the cavity in the parting line of the mold, as shown in Figure 4.8.

4.3 Basic Setting of the Mold Clamping Force

This sampling stage does not require a precise knowledge of the level of injection pressure, holding pressure or cavity pressure. However, the "mold sampling specialist" must be able to set a sufficient mold clamping force to facilitate the filling study presented in Chapter 6 "Filling study, holding pressure and mold clamping force". From the specific clamping force as a guide value and the projected area of the molded part, the theoretically required clamping force F_C can be determined approximately. The equation for this is as follows:

$$F_C = \text{Specific clamping force}\left(\frac{\text{kN}}{\text{cm}^2}\right) \times \text{Projected area}\left(\text{cm}^2\right) \times \text{Number of molded parts} \quad (4.1)$$

Table 4.1 shows guide values for the specific clamping forces of amorphous and semi-crystalline thermoplastics.

Table 4.1 Empirical Guide Values for Specific Clamping Forces

Plastic	Standard values for specific clamping force (kN/cm^2)	Empirical values for cavity pressure (bar)
Amorphous thermoplastics		
PS	1.5–3.5	150–350
SB	2.0–4.0	200–400
SAN	2.5–4.5	250–450
ABS	3.0–5.0	300–500
PC/ABS	3.0–5.5	300–550
PVC rigid	2.5–5.0	250–500
PVC plasticized	1.5–3.0	150–300
CA	2.5–4.5	250–450
CAB	2.5–4.5	250–450
CP	2.0–3.5	200–350
PMMA	3.5–5.5	350–550
PPE (mod.)	3.5–6.0	350–600
PPO (mod.)	3.5–6.0	350–600
PC	3.5–6.5	350–650
PSU/PES	4.0–6.0	400–600
PAR	3.5–6.5	350–650
PEI	4.5–6.5	450–650
PAI	4.5–7.5	450–750
Semi-crystalline thermoplastics		
PE-HD	2.0–6.0	200–600
PE-LD	2.0–6.0	200–600
PP	3.0–6.5	300–650

Table 4.1 Empirical Guide Values for Specific Clamping Forces (*continued*)

Plastic	Standard values for specific clamping force (kN/cm^2)	Empirical values for cavity pressure (bar)
PA 4.6	4.5–7.5	450–750
PA 6	3.5–5.5	350–550
PA 6.6	4.5–7.5	450–750
PA 6.10	3.0–5.0	300–500
PA 11, PA 12	3.5–4.5	350–450
PA amorphous	3.5–4.5	350–450
POM	5.5–10.5	550–1050
PET	4.5–7.5	450–750
PBT	4.0–7.0	400–700
PPS	3.5–6.5	350–650
PAA	3.0–7.0	300–700
LCP	3.0–8.0	300–800

PRACTICAL EXAMPLE:

Given:

- Molded part: Housing cover
- Plastic: PC/ABS
- Mold cavities: 1
- Projected area: 535.8 cm^2
- From Table 4.1, a specific clamping force for PC/ABS of 5.5 kN was selected.

Sought:

Provisional theoretical clamping force in kN

Solution:

Step 1:

Using Equation 4.1:

$$F_C = \text{Specific clamping force} \left(\frac{kN}{cm^2}\right) \text{ x Projected area } (cm^2) \text{ x Number of molded parts}$$

Step 2:

Substituting into Equation 4.1 :

$$F_C = 5.5\left(\frac{kN}{cm^2}\right) \text{ x } 535.8\left(cm^2\right) \text{ x } 1 = 2946.9\ kN$$

Answer:

A provisional clamping force of 2946.9 kN or approx. 290 metric tons is required.

Another reliable method for determining the preliminary clamping force is that of simulations. In many instances, the injection molded parts are already accompanied and optimized during the development and design phase by an injection molding simulation program. In this case, it is possible to display the required clamping force (Figure 4.10) for the housing cover shown in the practical example and to use it for the injection molding process.

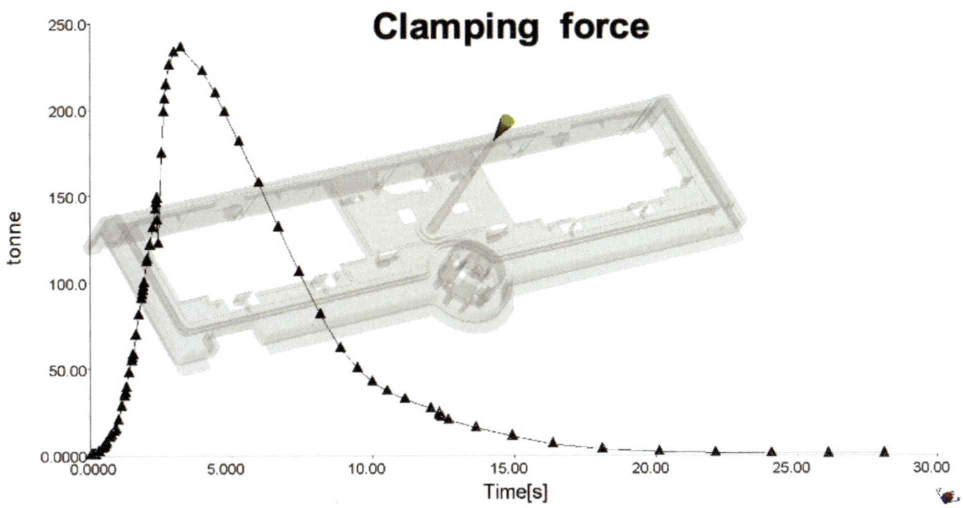

Figure 4.10 Preliminary clamping force obtained via an injection molding simulation (image source: Gentner Kunststoffverarbeitung GmbH)

The subsequent filling study (see also Chapter 6 "Filling study, holding pressure and mold clamping force") can be started with the clamping force calculated by the user or taken from the simulation. The clamping force will be discussed in detail later in Chapter 6 "Filling study, holding pressure and mold locking force". All important background information and an optimization strategy are presented there.

4.4 Mold Temperature Control

In this section, I would like to draw attention to the importance and difficulties of mold temperature control. Alongside the cavity pressure curve and the melt temperature, mold temperature control is the most important process parameter governing process reliability, dimensional accuracy and the visual quality of the molded part.

In many cases, no further attention is paid to mold temperature control once the temperature control hoses have been connected, usually without cooling plans.

As a result, a large part of the necessary process reliability and the required economic efficiency in the form of cycle time has already been lost at this stage of the molding process.

REMEMBER:

In practice, well over 50% of all molded part defects are due to incorrect thermal management and consequently uneven temperature distribution in the mold.

The goal of the "mold sampling specialist" should not only be to correctly set the mold temperature, but also to achieve an even temperature distribution in the injection mold.

In the following sections, I would like to show you all the necessary details along with further knowledge on the subject of mold temperature control, such as:

- Finding the optimum mold temperature required
- The effects of a mold temperature that is too low or too high for the thermoplastics being processed
- The possible error indicators that lead to uneven temperature distribution in the mold
- The causes of thermal faults in the injection mold
- Checking the flow rate of the temperature control medium

4.4.1 Mold Temperature

For this purpose, the "mold sampling specialist" requires the processing data sheet, which is obtained from the raw material manufacturer (an example is shown in Figure 4.11) and is already included in the sampling documents. This data sheet contains all the important or recommended values and processing instructions of the raw material manufacturer.

Bayblend® T85 XF RE

Property	Test Condition	Unit	Standard	typical Value
Recommended processing and drying conditions				
Melt temperatures		°C	-	260 - 280
Standard Melt temperature		°C	-	270
Barrel Temperatures - Rear		°C	-	230 - 240
Barrel Temperatures - Middle		°C	-	235 - 245
Barrel Temperatures - Front		°C	-	240 - 270
Barrel Temperatures - Nozzle		°C	-	265 - 275
Mold Temperatures		°C	-	70 - 90
Hold Pressure (% of injection pressure)		%	-	50 - 75
Plastic Back Pressure (specific)		bar	-	50 - 150
Peripheral Screw Speed		m/s	-	0.05 - 0.2
Shot-to-Cylinder Size		%	-	30 - 70
Dry Air Drying Temperature		°C	-	95 - 110
Dry Air Drying Time		h	-	4
Moisture Content max. (%)		%	-	<= 0,02
Vent Depth		mm	-	0.025 - 0.075

C These property characteristics are taken from the CAMPUS plastics data bank and are based on the international catalogue of basic data for plastics according to ISO 10350.

Impact properties: N = non-break, P = partial break, C = complete break

Figure 4.11 Excerpt from a processing data sheet supplied by a raw material manufacturer (image source: Covestro)

If you do not have the detailed processing instructions from the raw material manufacturer, you can use the following recommended guide values for mold temperature given in Table 4.2.

Table 4.2 Recommended Guide Values for the Mold Temperature

Plastic	Mold temperature (°C)	Plastic	Mold temperature (°C)
PE-LD	15–60	ASA	40–80
PE-HD	20–60	PMMA	50–90
PP	10–90	POM	60–100
PVC plasticized	10–60	CA	40–80
PVC rigid	20–60	CAB	40–80
PS	10–70	PA 6	50–90
S/B	10–70	PA 6 GF30	80–120
SAN	40–80	PA 66	50–90

Table 4.2 Recommended Guide Values for the Mold Temperature (*continued*)

Plastic	Mold temperature (°C)	Plastic	Mold temperature (°C)
ABS	50–80	PA 12	30–100
PC	80–100	PC/ABS	70–100
PET	130–140	PC/ABS GF20	80–90
PBT	40–100	PC/ASA	60–90
PSU	120–160	PC/PBT	60–80

You can see that, in practice, all recommended mold temperatures usually range very widely from the minimum to the maximum recommended values. This raises a legitimate question:

What should the "mold sampling specialist" set?

The following questions must be clarified before you can make the right decision during the sampling process:

- Which thermoplastic is to be processed?
- What effect does the mold temperature have on the thermoplastic being processed?
- What type of molded part is to be produced? A technical precision part (e. g. gearwheel) or a mass-produced item (e. g. packaging)?
- What requirements are placed on the molded part in terms of surface finish or surface quality?
- What requirements are placed on the molded part in terms of dimensional accuracy?
- What cycle time is to be adhered to or is specified?

4.4.1.1 Amorphous or Semi-Crystalline Thermoplastics

The amorphous thermoplastics (Figure 4.12) consist of long-chain molecules that intertwine like a ball of wool when they are formed. Due to their asymmetrical structure, they do not crystallize and are crystal-clear in their natural form if they have not been modified or colored. The application areas of amorphous thermoplastics lie below the glass transition temperature (T_g).

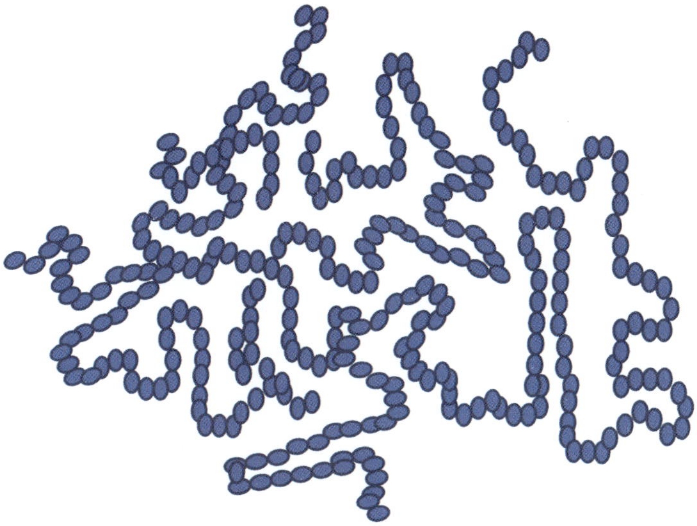

Figure 4.12 Structure of amorphous thermoplastics

If the temperature of the plastic melt is below T_g, the plastic part has solidified in the mold and can be demolded [1].

Examples of amorphous thermoplastics are ABS, PC, PMMA, PS, and PVC.

Semi-crystalline thermoplastics (Figure 4.13) have, as the name suggests, semi-ordered molecular regions called crystalline regions. Between the crystalline regions, there are amorphous molecular regions. Transparency decreases with increase in crystallinity, and so semi-crystalline thermoplastics are milky or opaque in their natural state, due to their dense arrangement. When cooling in the injection mold, the plastic melt crystallizes again. The density in the crystalline areas increases substantially and the volume decreases. This results in high shrinkage and a high tendency for the part to warp. In some cases, crystallization is still incomplete upon removal from the injection mold. This results in post-crystallization, which can take several weeks for complex part geometries. The application temperature ranges lie between the glass transition temperature (T_g) and the crystallite melting temperature (T_m). Their processing options are the same as for amorphous thermoplastics, but the cooling conditions and mold temperature have a major influence on the part properties, due to the different crystallinity and subsequent post-crystallization [1].

Examples of semi-crystalline thermoplastics are PE, PP, POM, and PTFE.

Particularly with semi-crystalline thermoplastics, care must be taken to ensure that crystallization in the mold is complete by the time the molded part is demolded. This requires a high mold temperature and therefore a higher processing shrinkage, followed by a slower cooling phase down to the recommended demolding temperature. Otherwise, increased post-crystallization will occur in the semi-crystalline materials, such as POM, PA, PP, PE or PBT, which in any event has a strong adverse impact on dimensional accuracy and gives rise to part distortion.

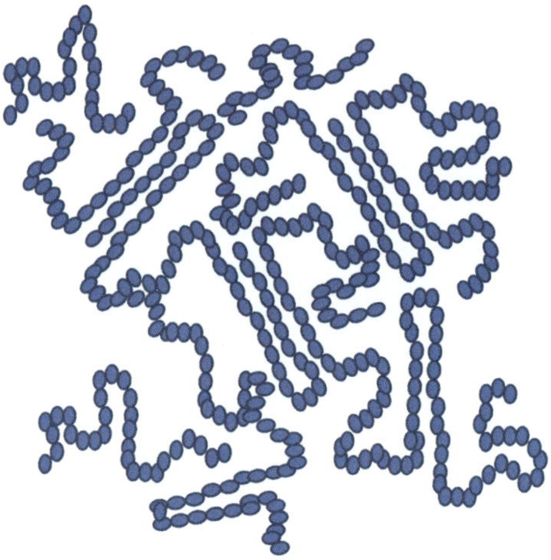

Figure 4.13 Structure of semi-crystalline thermoplastics

4.4.1.2 Effects of Mold Temperature on the Molded Part

In order to better understand the relationships between mold temperature and amorphous and semi-crystalline thermoplastics, Table 4.3 takes a closer look at the effects of too low and too high a temperature.

When the mold temperature is being selected, the thermoplastic to be processed and the associated effects (Table 4.3) must always serve as the basis for deciding between higher quality, such as appearance and dimensional accuracy, on one hand, and lower quality and consequently a more favorable unit price for the molded part, on the other.

The starting assumption here should be "only as good as necessary, not as good as possible" in order that a compromise may be reached between a partially optimized temperature and an economically viable solution.

NOTE:

An empirical value often quoted in practice and in training courses states that an increase in mold temperature of just 1 °C increases the required residual cooling time by up to 2%.

ENERGY EFFICIENCY:

For purely energy-related and economic reasons, always start with the lowest mold temperature recommended by the raw material manufacturer and only increase this if necessary.

Table 4.3 Effects of Mold Temperatures on Amorphous and Semi-Crystalline Thermoplastics

High mold temperatures mean:	
For amorphous thermoplastics:	**For semi-crystalline thermoplastics:**
A lower pressure requirement, as the flow properties of the plastic melt are improved	
A better surface finish or quality of the molded part	
Molded parts with fewer orientations, resulting in less stress	
Lower residual stresses in the molded part	
A favorable pressure effect exerted by the holding pressure in the molded part	
This can result in molded part dimensions that are too small	
An extension of the injection molding cycle, due to a longer residual cooling time	
The influence on processing and post-shrinkage is less	More crystallization, thus higher processing shrinkage with subsequent low post-shrinkage and consequently hardly any subsequent part distortion
A longer cooling phase until the demolding temperature is reached	A longer cooling phase for the required crystallization process until the demolding temperature is reached
Low mold temperatures mean:	
For amorphous thermoplastics:	**For semi-crystalline thermoplastics:**
A possible deterioration in the surface finish or quality, e. g. due to matt surfaces, vinyl record effects, visible flow lines and weld lines	
A non-optimal pressure effect exerted by the holding pressure on the molded part, due to areas of the molded part that freeze quickly, for example	
Possible condensation in the mold if the temperature difference between the mold and melt temperature is too high	
High residual stress in the molded part	
Possible cold displacements in the molded part	
Increased pressure requirement during injection	
More orientations in the molded par	
A shorter cooling phase until the demolding temperature is reached	
Shortening of the injection molding cycle	
The influence on processing and post-shrinkage is minimal	Reduced crystallization, resulting in low processing shrinkage with subsequent high post-shrinkage and consequently greater part distortion

4.4.2 Even Temperature Distribution in the Mold

In this section, I would like to emphasize the importance of an even temperature distribution in the mold. To this end, I will present the knowledge needed for quickly, efficiently and purposefully identifying and avoiding the causes of thermal faults in the injection mold in the subsequent mold sampling analysis (Chapter 7 "Mold sampling analysis of the basic setting").

Due to the ever-increasing complexity of molds, it is becoming increasingly difficult in practise to create optimum temperature control conditions. Some of the reasons for this can already be found in the mold design. Due to complex mold structures, some mold areas with different wall thicknesses do not allow ideal conformal cooling.

The resulting different distances between the cooling channels and the molded part result in uneven heat dissipation and consequently different temperature ranges in the mold, which leads to uneven temperature distribution. Such design problems cannot be sustainably corrected either with a high level of technical expertise or through consistent optimization trials.

Another reason in practice is that mold temperature systems are connected up to the best of one's knowledge and belief during a first sampling, but no further attention is paid to them subsequently. As a result, possible thermal problem areas in the mold can lead to considerable process problems during subsequent series production. An uneven temperature distribution affects the following molded part properties:

- Dimensional accuracy
- Different warping and shrinkage behavior in the molded part
- A different plastic structure (crystallization behavior)
- Weight of the molded part
- Different residual stresses in the part
- Fluctuations in mechanical properties
- Differences in surface quality and properties

NOTE:

In order to achieve a sufficiently even temperature distribution in the mold, a temperature deviation of ±5 °C should not be exceeded within the cavity area. Higher deviations generally result in warpage, surface defects, dimensional problems and longer residual cooling times.

4.4.2.1 Causes and Consequences of Uneven Temperature Distribution in the Injection Mold

The goal is to investigate and optimize the optimum temperature distribution in the mold during sampling. For this purpose, it is of decisive advantage to ask the following questions:

- Where can thermal problems occur in the injection mold?
- What is the cause of the thermal problems?
- What effects do these thermal problems have on the molded part?

Table 4.4 provides you with practical answers to these questions.

All causes and consequences of an uneven temperature distribution in fault categories with possible fault sources, fault causes and their effects on the molded part are dealt with in detail in the later part of this book.

Table 4.4 Causes and Consequences of Uneven Temperature Distribution [2]

Fault category	Fault source	Consequence of the fault
Improper connection of the mold temperature tubes	The temperature-control hoses are too long:  	Pressure loss in the temperature system Low flow values of the temperature control medium Consequently, poorer heat dissipation
	Kinked hoses: 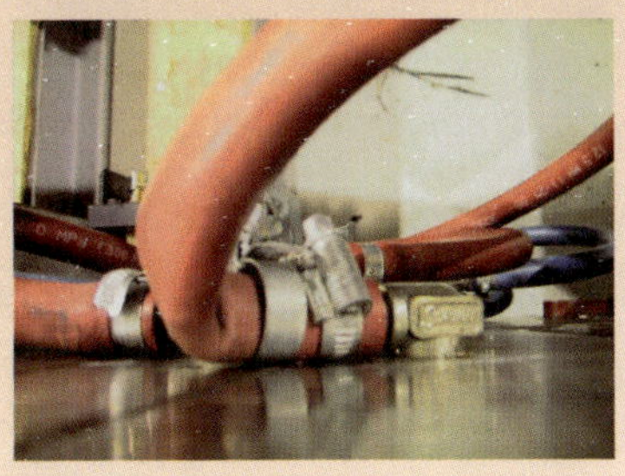	

Table 4.4 Causes and Consequences of Uneven Temperature Distribution [2] (*continued*)

Fault category	Fault source	Consequence of the fault
Technical condition of the temperature control and cooling units	Limescale and rust deposits in the lines: 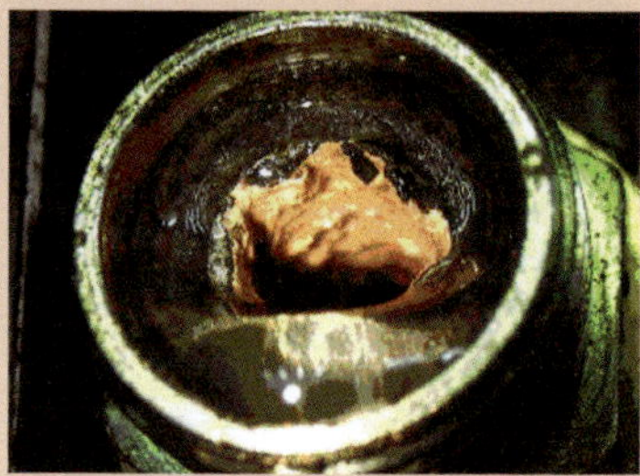	Pressure loss in the temperature system Low flow values of the temperature control medium Consequently, poorer heat dissipation
Hot runner system	Inadequate insulation of the mold: 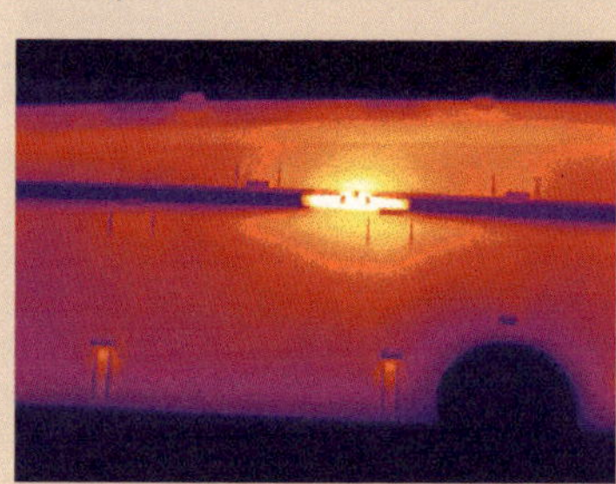	Strong heat radiation in the mold Creation of hot spots Uneven temperature level in the mold Excessive temperature deviations on the mold wall
	Asymmetrical heating: 	

Table 4.4 Causes and Consequences of Uneven Temperature Distribution [2] (*continued*)

Fault category	Fault source	Consequence of the fault
Temperature-control channels layout	Asymmetrical temperature-control channels layout in the mold: 	Uneven heat dissipation in the mold Excessive temperature deviations on the mold wall Uneven heat dissipation in the mold Excessive temperature deviations on the mold wall Hot spots
	Non-existent or deficient areas: 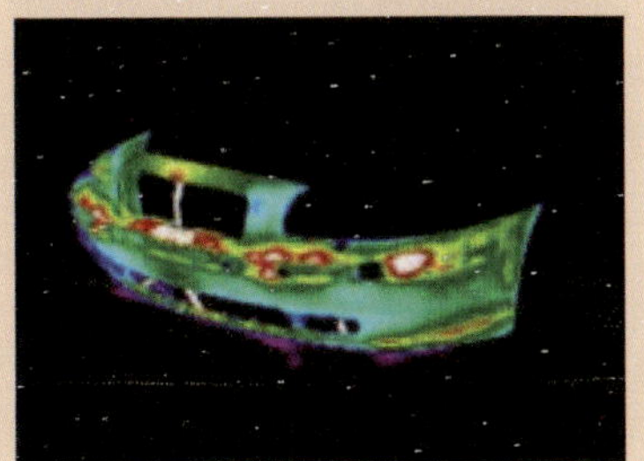	
Effects of these fault categories on the molded part:		
▪ Dimensional inaccuracy ▪ Variations in mechanical properties ▪ Different warping and shrinkage behavior ▪ Differences in surface quality and properties ▪ Different plastic structure (crystallization behavior); very long residual cooling time or cycle time		

image sources: GTT Willi Steinko GmbH

4.4.3 Checking the Flow Rate of the Temperature Control Medium

The flow rate (l/min) of the temperature control medium has a direct influence on the heat transfer in the temperature control channel of the injection mold. This means that the lower the flow rate, the poorer is the heat transfer and the greater the temperature difference between the temperature control channel and the heat transfer medium or temperature control medium.

For early detection of possible creeping reduction in the flow rate during subsequent series production as a result of

- Contamination and deposits of organic and inorganic substances in the temperature control system, e. g. limescale deposits or rust
- Tube kinks or crushing of pipework, and
- Pump wear (reduction in pump performance)

it is essential to record and document the actual flow rate value for each individual temperature control circuit during sampling with the aid of an external flow meter (Figure 4.14) or an integrated flow measurement on the injection molding machine (Figure 4.15).

Figure 4.14 Example of a modern flow meter (image source: Engel GmbH)

Figure 4.15 Example of integrated flow measurement at the injection molding machine (image source: Engel GmbH)

The measured flow rate value is a critical starting point when it comes to quality fluctuations and part flaws in subsequent series production. A further flow rate measurement performed in series production and serving as a reference value can very quickly reveal whether a thermal problem in the mold is possibly the reason for the fluctuations in quality.

PRACTICAL TIP:

- A flow measurement should be performed regularly, ideally for every mold change in subsequent series production, and compared with the flow value from the sampling.
- To avoid corrosion and deposits, the temperature control channels should always be cleaned of the temperature control medium, e.g. by purging with compressed air, even after the first sampling in the mold. This should be a standard procedure for employees during subsequent series production.

In the subsequent mold sampling analysis (Chapter 7 "Mold sampling analysis of the basic setting"), the amount of heat Q that must be introduced into the mold during the production of your molded part and dissipated again via the temperature control medium is checked. This allows you to calculate the theoretical water requirement and compare it with the actual flow rate in practice.

4.4.4 Heating Both Mold Halves Together

After setting the clamping unit, the mold halves (nozzle side and ejector side) should be carefully closed together in set-up mode, especially in the case of molds operated at higher temperatures.

This procedure has the following advantages:

- A uniform heating ratio between the nozzle side and ejector side
- Less heat loss, because the mold is closed
- With hot runner systems: heat transfer also takes place on the ejector side, which in turn favors uniform heating
- It is more gentle on the mold, as uniform thermal expansion occurs in the mold
- The mold can therefore heat up optimally during subsequent set-up of the plasticizing unit

NOTE:

If high mold temperatures are set, the mold must be re-zeroed after the heating phase, due to the thermal expansion of the mold in set-up mode. You should also check the functioning of the set mold protection mechanism.

Milestone: "Basic setting of the clamping unit"

- You have completed block 3 on the sample checklist
- The injection mold has been correctly set up on the machine
- The mold temperature has been set correctly
- The strokes, speeds and times of the clamping unit have been optimally set and checked
- The adjustment of the clamping unit is complete

References

[1] Baur, E., Harsch, G., Moneke, M: Werkstoff-Führer Kunststoffe, 11th edition, Carl Hanser Verlag, Munich, 2019

[2] GTT Willi Steinko GmbH

5 Basic Setting of the Plasticizing Unit

Setting of the clamping unit is followed by setting of the plasticizing unit. The goal here is to find the optimal basic setting for the molded part. The following sections contain the necessary knowledge and many practical tips to help you determine the settings for the machine. An important aspect in this connection is energy-efficient and low-wear machine settings.

During the setting phase of the plasticizing unit, the "mold sampling specialist" systematically works through block 5 ("Basic setting of the plasticizing unit") of the mold sampling checklist (Figure 5.1).

Basic setting of the plasticizing unit				
Temperatures:		**Yes**	**No**	**Block 5**
16	Barrel temperature correctly selected and set?			
17	Barrel temperature profile matched to thermoplastic?			
18	Correct flange temperature selected and set?			
Plasticizing process:		**Yes**	**No**	
19	Required plasticizing volume or stroke determined and set?			
20	Plasticizing speed set?			
21	Screw back pressure and screw decompression set?			
Injection process:		**Yes**	**No**	
22	Injection pressure set?			
23	Injection speed and injection speed profile set?			
Cooling time and demolding temperature:		**Yes**	**No**	
24	Required demolding temperature determined?			
25	Theoretical residual cooling time calculated and set?			
Plasticizing unit:		**Yes**	**No**	
26	Nozzle contact point zeroed?			
27	Nozzle impression created and OK?			
28	Movements (e.g. lifting of the nozzle) set?			

Figure 5.1 Mold sampling checklist: Block 5

5.1 Setting the Barrel Temperatures

Barrel temperature is an important machine setting parameter that is crucial for quality-relevant properties, such as surface finish, strength and dimensional accuracy of the molded part. After the screw speed and screw back pressure, the set barrel temperature has the greatest influence on the plastic melt.

To set the correct barrel temperature, it is necessary to use the raw material manufacturer's processing data sheet for the plastic as a guide. The sample data sheet in Figure 5.2 shows the recommended range of processing temperatures for the plastic melt.

covestro

Bayblend® T85 XF RE

Property	Test Condition	Unit	Standard	typical Value
Recommended processing and drying conditions				
Melt temperatures		°C	-	260 - 280
Standard Melt temperature		°C	-	270
Barrel Temperatures - Rear		°C	-	230 - 240
Barrel Temperatures - Middle		°C	-	235 - 245
Barrel Temperatures - Front		°C	-	240 - 270
Barrel Temperatures - Nozzle		°C	-	265 - 275
Mold Temperatures		°C	-	70 - 90
Hold Pressure (% of injection pressure)		%	-	50 - 75
Plastic Back Pressure (specific)		bar	-	50 - 150
Peripheral Screw Speed		m/s	-	0.05 - 0.2
Shot-to-Cylinder Size		%	-	30 - 70
Dry Air Drying Temperature		°C	-	95 - 110
Dry Air Drying Time		h	-	4
Moisture Content max. (%)		%	-	<= 0.02
Vent Depth		mm	-	0.025 - 0.075

C These property characteristics are taken from the CAMPUS plastics data bank and are based on the international catalogue of basic data for plastics according to ISO 10350.

Impact properties: N = non-break, P = partial break, C = complete break

Figure 5.2 Sample data sheet supplied by the raw material manufacturer and showing the processing temperature (image source: Covestro)

If there are no processing instructions supplied by the raw material manufacturer, you can use the recommended melt temperatures presented in Table 5.1 .

Table 5.1 Recommended Melting Temperatures for Various Plastics

Plastic	Melt temperature (°C)	Plastic	Melt temperature (°C)
PE-LD	170–270	ASA	240–280
PE-HD	200–290	PMMA	190–270
PP	200–300	POM	185–230
PVC plasticized	160–190	CA	170–210
PVC rigid	170–210	CAB	180–220
PS	180–280	PA 6	250–290
S/B	190–280	PA 6 GF30	270–290
SAN	200–270	PA 66	270–300
ABS	210–270	PA 12	200–270
PC	280–320	PC/ABS	230–280
PET	260–280	PC/ABS GF20	240–280
PBT	250–275	PC/ASA	240–300
PSU	330–360	PC/PBT	260–280

However, the processing temperatures recommended by raw material manufacturers should not be equated or confused with the barrel temperature to be set on the machine. This refers to the recommended temperature range of the plastic melt required for optimally processing the plastic.

It should be noted that in addition to the set barrel temperature, the plastic melt is heated by the frictional heat generated during plasticizing.

This frictional heat is caused by shearing (internal friction) in the plastic melt. In practice, it is very often the case that the temperature of the injection-molded plastic melt can be 10 °C to 30 °C higher than the set barrel temperature. This additional heat introduced into the melt is a factor that should not be underestimated, as a higher melt temperature influences the shrinkage properties and can lead to dimensional deviations in the molded part.

In order to ensure that the recommended melt temperature of the raw material manufacturers in the injection molding process is attained, refer to the review and assessment of the plastic melt presented in Section 5.7 "Assessment of the plastic melt".

ENERGY EFFICIENCY:

Heating energy accounts for a large proportion of the power consumption in the injection molding process, so you should always start with the lowest processing temperature recommended by the material manufacturer.

5.1.1 Thermal Behavior of Amorphous and Semi-Crystalline Thermoplastics

Amorphous and semi-crystalline thermoplastics exhibit different thermal behavior as the temperature in the plasticizing barrel rises from point of material feed to the state at which the plastic is ready for processing. These differences are very important for further settings, such as the optimum barrel temperature profile.

The "Background knowledge" information box below contains important details on the thermal behavior of amorphous and semi-crystalline thermoplastics as the temperature rises.

BACKGROUND KNOWLEDGE:

Thermal behavior of amorphous thermoplastics:

Initially, the plastic is in a hard-elastic state. In this state, the thermal movements of the molecules are so low that the thermoplastic offers high resistance to external forces. Consequently, the hard-elastic state is dimensionally stable and is therefore also the area of use for amorphous thermoplastics. In the hard-elastic state, it can be seen from Figure 5.3 that the tensile strength of the plastic below the softening range (T_g) is relatively high and the elongation at break is very low.

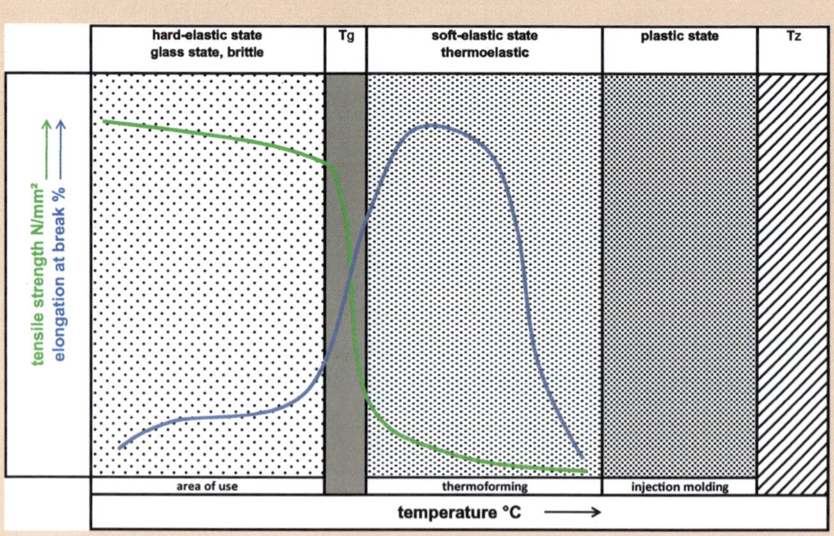

Figure 5.3 Ranges of state for amorphous thermoplastics

As amorphous thermoplastics do not have a melting point, they change from a hard-elastic to a soft-elastic state within a certain temperature range, called the glass transition or softening temperature (T_g). When the T_g is reached, the tensile strength drops sharply, due to the sudden onset of thermal movements of the plastic molecules. The elongation at break increases sharply in the T_g range and decreases in the soft-elastic state, due to the increasing plastic content of the polymer.

A further increase in temperature causes the thermoplastic to pass from the soft-elastic state to the plastic state. At this point, the thermal movement of the thermoplastic molecules is so high that the thermoplastic can be processed as a plastic melt by injection molding. However, if the plastic state range is exceeded owing to a further rise in temperature, the decomposition temperature range (T_z) of the thermoplastic is reached. The amorphous thermoplastic begins to decompose thermally. This takes the form of discoloration, bubble formation and gas development.

Thermal behavior of semi-crystalline thermoplastics:

The starting point for semi-crystalline thermoplastics is the hard-elastic state. Figure 5.4 shows that semi-crystalline thermoplastics, too, have a glass transition temperature. During the transition from the hard-elastic to the soft-elastic state, only the amorphous areas become soft-elastic when the glass or softening temperature (T_g) is reached. The remaining portion of semi-crystalline areas is still solid and dimensionally stable, due to the cohesion of the molecules.

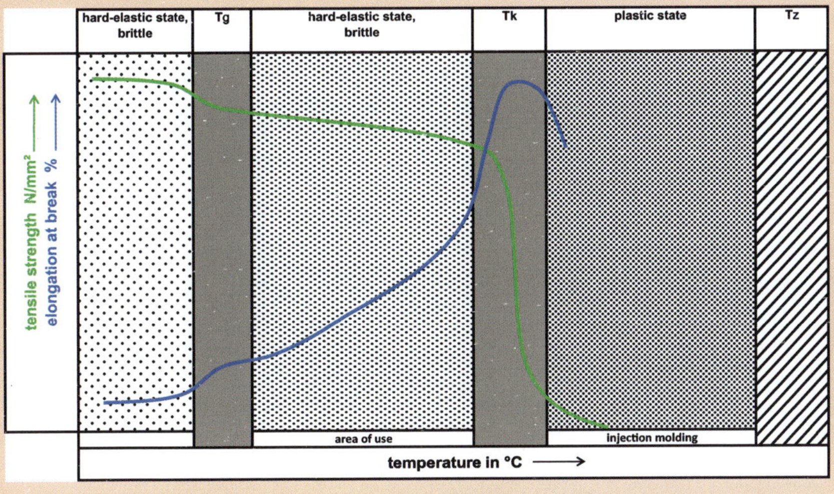

Figure 5.4 The different states of semi-crystalline thermoplastics

This property is characterized by a slight discontinuity in the tensile strength and elongation curves. After a further rise in temperature, the soft-elastic state changes to the thermoplastic state when the crystallite melting temperature (T_K) is reached. The area of use for semi-crystalline thermoplastics lies in the soft-elastic state between T_g and T_K. It is only at T_K that the mechanical strength decreases abruptly and that moldability increases. The thermal movement of the molecules is so great here that the melt can be processed by injection molding.

Consequently, the semi-crystalline thermoplastics have a definite melting point. If the plastic state range is exceeded by a further rise in temperature, the decomposition temperature range (T_z) of the plastic is reached.

Here, the semi-crystalline thermoplastic begins to decompose thermally. This takes the form of discoloration, bubble formation and gas development.

5.1.2 Effects of the Melt Temperature on the Molded Part and the Injection Molding Process

The melt temperature has many effects on the process and the molded part. These must be identified in order that they may be used for the subsequent molded part or for optimizing the injection process. To this end, a matrix has been compiled of the possible effects of a high and low melt temperature on the molded part and the injection molding process.

Table 5.2 hows a large number of effects, some of which interact with each other. The aim should be to find the optimum compromise for the required molded part quality, cost-effectiveness and energy-efficiency on the basis of these individual effects.

Table 5.2 Influences of the Melt Temperature on the Molded Part and on the Injection Molding Process

Effects of the melt temperature on the injection molding process:	
High melt temperature:	**Low melt temperature:**
Longer holding pressure time possible	Shorter holding pressure time
Better distribution of the holding pressure possible	Worse distribution of the holding pressure
Higher shrinkage potential	Lower shrinkage potential
More crystallinity	Less crystallinity
Improved surfaces, e. g. visibility of flow and weld lines	Possible visible surface defects
Fewer residual stresses	More residual stresses
Fewer molecular orientations	More molecular orientations
Larger part dimensions	Smaller part dimensions
Longer necessary residual cooling time	Shorter necessary residual cooling time

Table 5.2 Influences of the Melt Temperature on the Molded Part and on the Injection Molding Process (*continued*)

Effects of the melt temperature on the injection molding process:	
High melting temperature:	**Low melting temperature:**
Lower injection pressure required	Higher injection pressure required
Thermal load on the melt	Gentle processing of the melt
▪ Reduced melt viscosity ▪ **Amorphous thermoplastics:** For every 1 °C temperature rise, 10–15% change in viscosity ▪ **Semi-crystalline thermoplastics:** ▪ For every 1 °C temperature rise, approx. 3% change in viscosity	Higher melt viscosity
Longer flow paths feasible	Shorter flow paths
Longer cycle time	Shorter cycle time
Higher energy consumption	Lower energy consumption

5.1.3 Setting the Barrel Temperature Profile

Following the selected barrel temperature of the plastic, the correct barrel temperature profile is crucial for gentle and homogeneous melting of the plastic. As already shown, amorphous and semi-crystalline thermoplastics exhibit different thermal behavior. Accordingly, the barrel temperature profile of amorphous and semi-crystalline thermoplastics must be adapted differently.

5.1.3.1 Temperature Profile for Amorphous Thermoplastics

Since the tensile strength of amorphous thermoplastics drops very quickly as the temperature rises after the T_g (glass transition temperature) has been reached, and rapid melting occurs as a result, it is advisable to set a decreasing barrel temperature profile (Figure 5.5) of 20–30 °C from the barrel nozzle to the first heating zone (HZ 1).

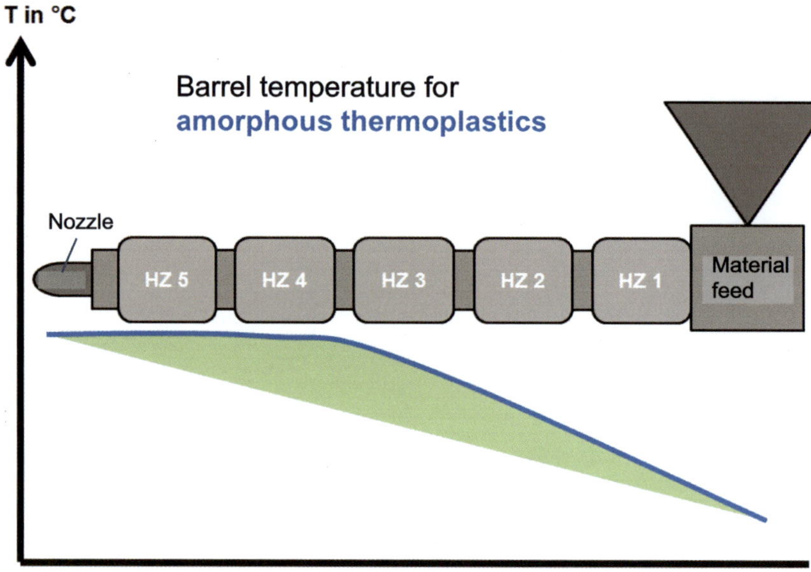

Figure 5.5 Recommended barrel temperature profile for amorphous thermoplastics (HZ = heating zone)

Table 5.3 shows some examples of barrel temperature profiles for amorphous thermoplastics.

Table 5.3 Barrel Temperature Profiles for Amorphous Thermoplastics in °C

Plastic	Nozzle	HZ 5	HZ 4	HZ 3	HZ 2	HZ 1
ABS	240	240	235	230	220	215
PC	290	290	285	280	275	270
PMMA	250	250	250	240	230	230
SAN	240	240	235	230	220	220
ASA	250	250	245	240	230	230
PS	240	240	235	230	220	220

5.1.3.2 Temperature Profile for Semi-Crystalline Thermoplastics

The tensile strength of semi-crystalline thermoplastics decreases more slowly as the temperature rises and only drops very quickly from the soft-elastic state to the plastic state when the T_m (crystallite melting temperature) is reached. Therefore, the temperature profile (Figure 5.6) from the barrel nozzle to the first barrel heating zone

(HZ 1) should be kept constant to slightly lower. In the case of polyamide 66, for example, high temperatures are used in the first heating zone (HZ 1) in order to achieve more gentle and homogeneous melting.

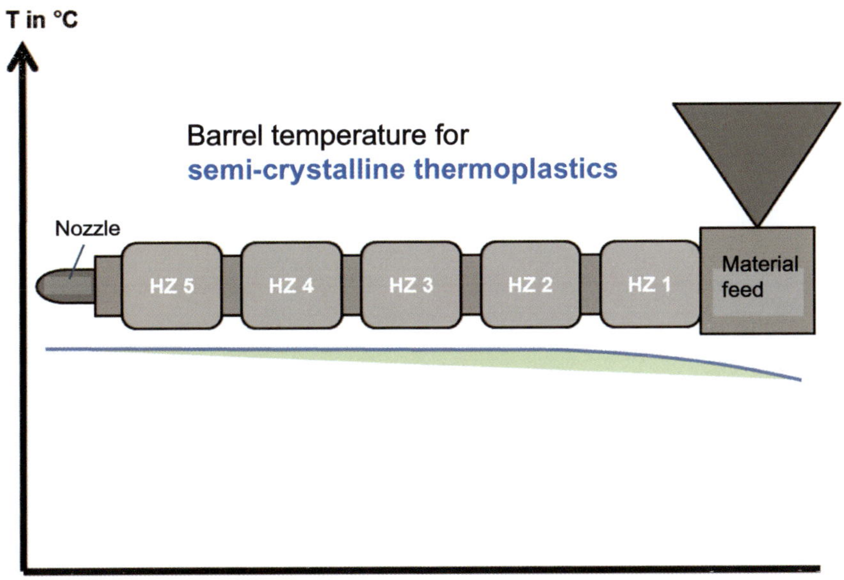

Figure 5.6 Recommended barrel temperature profile for semi-crystalline thermoplastics (HZ = heating zone)

Some examples of barrel temperature profiles for semi-crystalline thermoplastics are listed in Table 5.4.

Table 5.4 Barrel Temperature Profiles for Semi-Crystalline Thermoplastics in °C

Plastic	Nozzle	HZ 5	HZ 4	HZ 3	HZ 2	HZ 1
PA 6	270	270	270	270	260	260
PA 66	290	290	290	290	285	280
PP	230	230	230	225	220	220
PE	220	220	220	215	210	210
POM	200	200	200	200	190	190
PBT	260	260	260	255	250	250

5.1.3.3 Temperature Profile for Fiber-Reinforced Plastics, High Crystallinity and High Throughputs

Another very important temperature profile (Figure 5.7) is used for fiber-reinforced plastics, high melt crystallinity and high required material throughputs. Here, a slightly increasing temperature profile from the barrel nozzle to the first barrel heating zone (HZ 1) of a maximum of 15–25 °C is used.

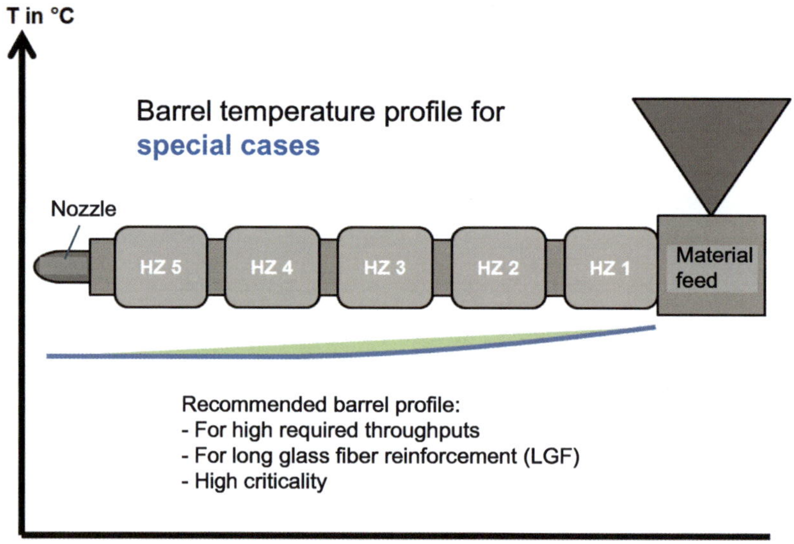

Figure 5.7 Recommended barrel temperature profile for special cases (HZ = heating zone)

The following "Background knowledge" information box contains information on the barrel temperature profile for fiber-reinforced, highly crystalline plastics and high material throughputs.

BACKGROUND KNOWLEDGE:

Fiber-reinforced plastics:

Nowadays, many components across all industries, such as front ends, vehicle dashboards, housings and functional components in the electrical sector, etc. are manufactured from glass-fiber-reinforced thermoplastics. The reason for this is that fiber-reinforced plastics possess enhanced mechanical properties, such as tensile, compressive, impact and flexural strength, compared with non-reinforced plastics. Short glass fibers (fiber length 0.2–0.5 mm) and long glass fibers (fiber length ≥ 10 mm) are used here. Long glass-fiber-reinforced plastics (LGF) are particularly critical, as the required fiber lengths need to reach the part during processing. The problem is that the fibers break and shorten during plasticizing and injection, due to high shear forces.

The fibers are exposed to high shear, especially at the start of plasticizing, as the melt (matrix) has not yet melted. To ensure gentle feeding of the fiber-reinforced material, a slightly increasing temperature profile (Figure 5.7) should therefore be used from the barrel nozzle to the first barrel heating zone (HZ 1).

High melt crystallinity:

If a material with a very high crystalline content has to be processed, higher temperatures and a longer melting phase are required until the T_m is reached. In order to avoid high frictional heat in the melt from the start of the feed, a slightly increasing temperature profile from the barrel nozzle to the first barrel heating zone (HZ 1) should be selected for gentle and homogeneous melting.

High material throughputs:

High throughputs (required injection volume and number of injection cycles per minute) of the material mean a shorter dwell time for the plastic melt in the barrel and therefore less time to obtain the thermoplastic in the required plastic processing state. It is therefore important to optimize the temperature profile (Figure 5.7) from the barrel nozzle to the first barrel heating zone (HZ 1). This has the advantage that less frictional heat due to shearing of the melt in HZ 1 and thus improved feed of the material can be realized.

5.1.4 Setting the Flange Temperature (Material Feed)

In this section, I would like to alert you to the importance of a correctly set flange temperature and explain the influences of this and the necessary process background knowledge. The flange temperature is an important process parameter that is very often given too little attention in practice. As a result, incorrectly set flange temperatures can be observed in many injection molding companies. This has its origins in the early days of injection molding, when the plastics of that time were processed in powder form, such as PVC, and were run with cold water ("hard flange-temperature control"). However, this type of temperature control has a negative effect on the injection molding of all the engineering plastics we process today, such as ABS, PA 6.6, PC, and PMMA. For these, the temperature of the flange area should be controlled with warm water ("soft flange-temperature control").

5.1.4.1 Effects of the Flange Temperature

The flange temperature plays a decisive role in the feed behavior and the feed stability of the plastic as well as the gentle plasticizing of the melt.

Table 5.5 provides a clear overview of the effects of the flange temperature on the injection molding process.

Table 5.5 Influences and Effects of the Flange Temperature

Flange setting:	Effects:
Flange temperature set too cold (e. g. for PA 6.6 with a flange temperature of 30 °C)	▪ Prolonged plasticizing time ▪ Uneven feeding of the plastic ▪ The plastic melts too late ▪ Poorer homogenization of the melt ▪ Greater frictional heat in the feed area of the screw
Excessive temperature differences between the flange temperature and the first heating zone (HZ 1) (e. g. for PA 6 with 280 °C in HZ 1 and a set flange temperature of 20 °C)	▪ Increased temperature-control requirement in HZ 1 ▪ Higher energy consumption for heating ▪ Possible formation of condensation in the feed area ▪ Moisture streaks on the molded part ▪ Material conveying problems, due to bridging in the feed area ▪ Premature melting of the plastic ▪ Possible air inclusions in the melt
Correctly set flange temperature (e. g. for PA 6 with a set flange temperature of 70 °C)	▪ Uniform feeding of the plastic ▪ Short plasticizing time ▪ Optimum melting of the plastic ▪ Gentle plasticizing process

Table 5.5 shows that not only the barrel temperature and the barrel temperature profile have a decisive influence on the plasticizing process, but that the flange temperature also plays an important role. For this reason, the "mold sampling specialist" should also watch out for a correctly controlled flange temperature.

5.1.4.2 Setting the Correct Flange Temperature

The "mold sampling specialist" must ask themselves the following questions in order to be able to select the correct flange temperature:

- What flange temperature is required?
- How do I get the correct flange temperature?

Firstly, I would like to give you an answer to the first question. The decisive factor here is whether you are processing an amorphous or a semi-crystalline thermoplastic. A semi-crystalline thermoplastic requires a longer melting phase and higher processing temperatures than an amorphous thermoplastic. Consequently, a higher flange temperature is required here, which favors earlier and gentler melting of the plastic. Another decisive factor for the flange temperature is whether the plastic to be processed, e. g. PA 6, PMMA, ABS, PC/ABS, and PET, needs to be pre-dried.

In this case, the warm plastic (e. g. PA 6 at 80 °C) is always run at a higher temperature to ensure an optimum thermal transition in the feed area.

NOTE:
All plastics that have to be pre-dried should have roughly the set drying temperature in the flange area as well.

The answer to the question as to where can I get the correct flange temperature can usually be found in the processing data sheets of the raw material manufacturers. In the event that no precise information or values are available, Table 5.6 provides an overview of some thermoplastics with the recommended flange temperatures.

Table 5.6 Recommended Flange Temperatures (°C)

Amorphous thermoplastics:			
PS	30–50	PVC	30–50
ABS	40–60	PC	70–90
PC/ABS	50–70	PMMA	60–80
SB	30–50	SAN	30–50
ASA	40–60	PC/ASA	50–70
PSU	45–65	PES	45–65
Semi-crystalline thermoplastics:			
PE	30–50	PP	30–50
POM	40–50	PA 6	60–90
PA 6.6	60–90	PET	50–70
PBT	50–70	LCP	45–65
Thermoplastic elastomers (TPEs):			
TPE-A, TPE-S, TPE-U, TPE-O, TPE-E			30–35

5.2 Setting the Plasticizing Process

The plasticizing process is one of the most important steps in the injection molding process. The task here is to achieve fast, gentle and homogeneous melting of the plastic while taking the energy efficiency in the barrel into account.

The following process parameters are decisive for the plasticizing process:

- Plasticizing stroke or volume
- Plasticizing speed (speed or circumferential speed of the screw)
- Screw back pressure
- Screw decompression (screw retraction)

In the following subsections, the necessary background knowledge, the influences and effects as well as a recommended procedure for setting these process parameters are explained in more detail.

5.2.1 Plasticizing Stroke or Volume

In practice, the plasticizing stroke (mm) and the plasticizing volume (cm^3) are generally used. In most cases, a suitable machine size for the molded part to be sampled is determined and selected in advance. Generally, however, the "mold sampling specialist" should ensure that the required plasticizing stroke (mm) does not utilize the maximum possible stroke of the plasticizing unit. In this case, the plasticizing stroke should be between one and three times the size of the screw diameter D (*1 D < plasticizing stroke > 3 D*). This means that, for a screw diameter of 35 mm, the optimum range of the plasticizing stroke is between 35 mm and 105 mm.

All the effects of a plasticizing stroke that is too short or too long are summarized in Table 5.7.

Table 5.7 Effects of a Non-Optimal Plasticizing Stroke

Plasticizing stroke too long > 3 *D*:
An uneven temperature profile along the length of the screw
Despite the high back pressure, the air cannot escape via the hopper in the direction of the filler opening
Air is not compressed until late in the compression zone, resulting in streaks on the molded part
Plasticizing stroke too short < 1 *D*:
If the plastic melt remains in the barrel for too long, there is a risk of thermal damage
Process fluctuations in the plasticizing stroke, because the resolution of the injection molding machine's stroke measuring system may not be sufficiently accurate
Possible fluctuations during injection, because the melt flows back into the screw flights in an undefined manner per shot
Insufficient melt homogeneity

5.2.1.1 Determining the Required Plasticizing Volume (cm^3)

The required plasticizing volume (cm^3) is determined from the shot weight (g) and the melt density (g/cm^3) of the plastic to be processed, also known as the output factor. As the density of plastics can vary considerably with the amount of heat introduced, the temperature of the environment into which the plastic is introduced plays a role. The density at processing temperature is lower than that of glass-hard plastic, due to the fact that the heat energy requires space.

The output factor takes into account the difference between melt and solid density.

REMEMBER:

Thermoplastics have a greater volume at melting temperature than at room temperature.

- For amorphous plastics, approx. 10% more volume is to be expected, as a result of volume change during heating
- For semi-crystalline plastics, approx. 15% more volume is to be expected, as a result of volume change during heating

For cold connections, the sprue weight must be added to the part weight. As there are generally no molded parts that can be weighed in the case of new molds, the weight information from the existing parts drawing can be used or the theoretical shot weight can be obtained from the responsible CAD designer. The required plastic density (g/cm^3) in the solid state can be found in the material data sheet of the respective raw material manufacturer. For subsequent work, you should always compile the required melt densities of your plastics in-house.

The required melt cushion (cm^3) is used to guarantee pressure transfer to the melt during the holding pressure time in order to compensate for volume shrinkage.

REMEMBER:

Difference between melt cushion and residual melt cushion:

The melt cushion is the plastic melt remaining in front of the screw tip (screw antechamber) after the injection phase. The residual melt cushion is the melt that remains in the antechamber after the holding pressure phase. This residual melt cushion serves as an important process monitoring parameter in downstream series production.

In the injection molding process, the residual melt cushion is subject to certain fluctuations. These are due to the fact that the same, reproducible conditions do not always prevail in the process, for example the screw position after plasticizing, the switchover point from injection to holding pressure and the closing behavior of the non-return valve. The extent of these fluctuations depends on the plastic to be processed and the screw diameter of the machine.

In practice, different empirical values are in circulation concerning the amount of residual melt cushion. There is information to the effect that the cushion should be kept smaller for thermally sensitive thermoplastics, such as POM, PC/ABS, and PVC, than for thermally insensitive thermoplastics, such as PS, PE, and PP. In principle, you should select the residual melt cushion so that a certain permitted fluctuation range in the process is compensated for and enough melt remains in front of the screw tip.

It is therefore advisable to start with a value of approx. 10mm residual melt cushion and, if necessary, to optimize and adjust it to the current injection molding process in the sampling phase (Chapter 6 “Filling study, holding pressure and mold clamping force”).

Table 5.8 shows the volume of the residual melt cushion at a distance of 10mm from the screw tip to the front end of the barrel as a function of the screw diameter.

Table 5.8 Volume of the Residual Melt Cushion at 10 mm as a Function of the Screw Diameter

Screw diameter (mm)	10mm Residual melt cushion (cm^3)	Screw diameter (mm)	10mm Residual melt cushion (cm^3)
12	1	60	28
15	1.7	70	38
18	3	80	50
25	5	90	64
30	7	100	79
35	10	110	95
40	13	120	112
45	16	130	183
50	20		

In practice, due to different machine makes, some machines work with stroke distances and some with volumes. However, all modern machines have an option for switching from strokes to volumes by means of an automatic internal conversion program.

I recommend working in volumes (cm^3) in the future, as this has the advantage that the values are easier to transfer to other machines and are more intelligible, because the volume (cm^3) is independent of the existing screw diameter, for example.

Equation 5.1 shows how the required plasticizing volume (cm³) is calculated:

$$\text{Plasticizing volume}\left(\text{cm}^3\right) = \left[\frac{\text{Shot weight (g)}}{[[\text{Output factor}]]\left(\frac{\text{g}}{\text{cm}^3}\right)}\right] + \text{Melt cushion}\left(\text{cm}^3\right) \quad (5.1)$$

Table 5.9 Guide Values for the Output Factor (g/cm³) of Selected Thermoplastics

Plastic	Output factor	Measured at melt temperature °C
ABS/PC	0.95	260
ABS	0.92	240
PA 66	0.96	280
PA 6	0.96	260
PA 6.6 GF 30	1.15	280
PC	1.04	300
PC GF 30	1.30	300
PBT	1.07	260
PBT GF 30	1.36	260
PC/PBT	1.04	265
PE-HD	0.72	220
PE-LD	0.70	220
PMMA	1.04	240
PET	1.10	260
POM	1.15	220
PP	0.73	220
PP 20% Talcum	0.85	230
PP 40% Talcum	0.98	230
PS	0.92	220
PVC	1.12	200
SAN	0.96	240

Table 5.9 Guide Values for the Output Factor (g/cm³) of Selected Thermoplastics (*continued*)

Plastic	Output factor	Measured at melt temperature °C
TPE-S	0.98	210
TPE-V	0.72	210
TPE-U	0.99	200

PRACTICAL EXAMPLE:

Given:

- Plastic: ABS
- Screw diameter: 40mm → Melt cushion at 10 mm: approx. 13 cm³
- Shot weight: 183 g
- Output factor: 0.92 g/cm³

Sought: Required plasticizing volume (cm³)

Solution:

$$\text{Plasticizing volume}\left(\text{cm}^3\right) = \left[\frac{183\ (\text{g})}{0.92\left(\frac{\text{g}}{\text{cm}^3}\right)}\right] + 13\left(\text{cm}^3\right) = 211.91\left(\text{cm}^3\right)$$

If you only want to work with strokes, it is important to convert the determined plasticizing volume as a function of the screw diameter. Table 5.10 can be used to quickly convert the plasticizing volume into a plasticizing stroke.

Table 5.10 Conversion of Plasticizing Volume (cm³) into Plasticizing Stroke (mm)

Screw diameter (mm)	1 cm³ Plasticizing volume corresponds to	1mm Plasticizing stroke corresponds to
12	8.85 mm	0.11 cm³
15	5.66 mm	0.17 cm³
18	3.93 mm	0.25 cm³
20	3.18 mm	0.31 cm³
22	2.63 mm	0.38 cm³
25	2.04 mm	0.49 cm³
30	1.41 mm	0.71 cm³

Table 5.10 Conversion of Plasticizing Volume (cm^3) into Plasticizing Stroke (mm) (*continued*)

Screw diameter (mm)	1 cm^3 Plasticizing volume corresponds to	1mm Plasticizing stroke corresponds to
35	1.04 mm	0.96 cm^3
40	0.80 mm	1.28 cm^3
45	0.63 mm	1.59 cm^3
50	0.51 mm	1.96 cm^3
55	0.42 mm	2.37 cm^3
60	0.35 mm	2.82 cm^3
65	0.30 mm	3.32 cm^3
70	0.26 mm	3.85 cm^3
80	0.20 mm	5.02 cm^3

Checking the set plasticizing volume (cm^3):

As an important test, the determined shot volume (cm^3) should be plasticized on the machine and ejected into the open.

The weight of the cooled plastic mass should be approximately the same as the theoretical weight according to the molded part drawing.

5.2.2 Plasticizing Speed

During the plastic melting process, approx. 30% of the heat is supplied by the electric heating bands and approx. 70% by the frictional heat generated by the rotating screw. The aim must therefore be to achieve that plasticizing speed which optimally plasticizes and homogenizes the plastic melt on one hand and does not damage it thermally or mechanically, due to excessive frictional heat, on the other.

5.2.2.1 Screw Speed and Screw Circumferential Speed

The plasticizing speed is specified via the screw speed (1/min) or the screw circumferential speed (m/min or m/s). The screw circumferential speed depends on the screw speed and the screw diameter (mm). In this context, I would like to explain some important practical details on the subject of screw speed and screw circumferential speed in the following “Background knowledge” information box.

BACKGROUND KNOWLEDGE:

Some injection molding companies still use older makes of machine, which in turn are equipped with older machine controls. It is always advisable to convert the screw speed into the screw circumferential speed, as this is the only way to compare the plasticizing effect on other machines with different screw diameters. If, for example, you were to compare the same speeds with different screw diameters, there would be considerable differences in frictional heat (friction) during the plasticizing process, which could have a negative effect on the melt quality.

With newer machines, the existing screw diameter is usually stored in the machine program so that the machine automatically calculates the required screw speed (1/min) after the screw circumferential speed (m/min or m/s) has been entered.

The required conversion of the screw speed (1/min) into the screw circumferential speed (m/s) can be found in Table 5.11 or calculated from the equation in the next practical example.

PRACTICAL EXAMPLE:

An injection molding screw with a diameter of 30 mm rotates at 200 revolutions per minute.

What is the circumferential speed of the screw during the plasticizing process?

Given:

$n = 200$ rpm

$d = 30$ mm

Sought: v_p

Solution:

$v_p = n * \pi * d$

$v_p = 200 * \pi * 30$ (mm/min) = 18.840 mm/min

= **0.31 m/s**

NOTE:

The screw circumferential speed should not be less than 0.06–0.08 m/s, as otherwise optimum homogenization of the melt cannot be guaranteed. It is also important to ensure that circumferential speeds of 0.33–0.66 m/s are maintained for thermoplastics, unless otherwise recommended by the raw material manufacturer, in order to avoid possible thermal damage to the melt.

Table 5.11 Conversion of Rotational Speed Into Circumferential Speed (m/s) [2

Rotational speed (1/min)	Screw diameter (mm)													
	18	25	30	35	40	45	50	60	70	80	90	100	110	120
20	0.02	0.03	0.03	0.04	0.04	0.05	0.05	0.06	0.07	0.084	0.094	0.105	0.12	0.13
40	0.04	0.05	0.06	0.07	0.08	0.09	0.10	0.13	0.15	0.17	0.19	0.21	0.23	0.25
60	0.06	0.08	0.09	0.11	0.13	0.14	0.16	0.19	0.22	0.25	0.28	0.31	0.37	0.38
80	0.08	0.10	0.13	0.15	0.17	0.19	0.21	0.25	0.29	0.34	0.38	0.42	0.46	0.50
100	0.09	0.13	0.16	0.18	0.21	0.24	0.26	0.31	0.37	0.42	0.47	0.52	0.58	0.63
120	0.11	0.16	0.19	0.22	0.25	0.28	0.31	0.38	0.44	0.50	0.57	0.63	0.69	0.75
140	0.13	0.18	0.22	0.26	0.29	0.33	0.37	0.44	0.51	0.59	0.66	0.73	0.81	0.88
160	0.15	0.21	0.25	0.29	0.34	0.38	0.42	0.50	0.59	0.67	0.75	0.84	0.92	1.01
180	0.17	0.24	0.28	0.33	0.38	0.42	0.47	0.57	0.66	0.75	0.85	0.94	1.04	1.13
200	0.19	0.26	0.31	0.37	0.42	0.47	0.52	0.63	0.73	0.84	0.94	1.05	1.15	1.26
220	0.21	0.29	0.35	0.40	0.46	0.52	0.58	0.69	0.81	0.92	1.04	1.15	1.27	1.38
240	0.23	0.31	0.38	0.44	0.50	0.57	0.63	0.75	0.88	1.01	1.13	1.26	1.38	1.51
260	0.25	0.34	0.41	0.48	0.54	0.61	0.68	0.82	0.95	1.09	1.23	1.36	1.50	
280	0.26	0.37	0.44	0.51	0.59	0.66	0.73	0.88	1.03	1.17	1.32	1.47		
300	0.28	0.39	0.47	0.55	0.63	0.71	0.79	0.94	1.10	1.26	1.41	1.57		
320	0.30	0.42	0.50	0.59	0.67	0.75	0.84	1.01	1.17	1.34	1.51			
340	0.32	0.45	0.53	0.62	0.71	0.80	0.89	1.07	1.25	1.42				
360	0.34	0.47	0.57	0.66	0.75	0.85	0.94	1.13	1.32	1.51				
380	0.36	0.50	0.60	0.70	0.80	0.90	0.99	1.19	1.39					
400	0.38	0.52	0.63	0.73	0.84	0.94	1.05	1.26	1.47					
420	0.40	0.55	0.66	0.77	0.88	0.99	1.10	1.32	1.54					
440	0.41	0.58	0.69	0.81	0.92	1.04	1.15	1.38						
460	0.34	0.60	0.72	0.84	0.96	1.08	1.20	1.45						
480	0.45	0.63	0.75	0.88	1.01	1.13	1.26	1.51						
500	0.47	0.65	0.79	0.92	1.05	1.18	1.13							

5.2.2.2 Effects of the Plasticizing Speed

When setting the correct plasticizing speed, you should know all the effects on the plastic melt and the injection molding process. Table 5.12 clearly shows what a plasticizing speed that is set too high or too low can mean for the plastic melt or the injection molding process.

Table 5.12 Effects of the Plasticizing Speed

Plasticizing speed too high:
Increased frictional heat (friction), which leads to thermal and mechanical damage to the plastic (reduction in the mechanical strength of the molded part)
Deterioration of melt homogeneity, due to non-optimally melted plastic melt
Increased wear on the plasticizing screw and non-return valve in the case of reinforced plastics, e. g. with glass or carbon fibers
Increased shear stress on the reinforcing fibers in the plastic melt (fiber breakage; shortening of the fibers and thus reduction in mechanical strength)
An increased energy requirement
Plasticizing speed too low:
An extended injection molding cycle

ENERGY EFFICIENCY:

As already described in Section 1.4 "Energy efficiency begins with sampling", the plasticizing phase is one of the most energy-intensive intervals in the injection molding process. A plasticizing speed that is optimally adjusted to the plastic and the process offers increased potential for energy savings.

5.2.2.3 Procedure for the Basic Setting of the Plasticizing Speed

The recommended screw speed can be found in the raw material manufacturer's processing data sheets. If no processing data is available, it is possible to request this from the raw material manufacturer's application technology department.

The recommended screw speed provided by the raw material manufacturer is usually specified with a minimum and a maximum screw speed. As no exact residual cooling time is known at this stage of the sampling phase, you should start with the maximum recommended speed and optimize it downward during the subsequent mold sampling analysis (Chapter 7 "Mold sampling analysis of the basic setting"). In principle, the setting should be selected so that the plasticizing time is complete 10–20% before the required residual cooling time has elapsed.

5.2.3 Screw Back Pressure

Unfortunately, it is often the case in practice that no special attention is paid to the back pressure. In spite of the back pressure recommendations made by the raw material manufacturers, settings are simply made on the spur of the moment without further consideration or subsequent checking. In this section, I would like to draw your attention to screw back pressure and provide valuable background information and suggestions.

The screw back pressure is an important process parameter, as it is crucial for the homogeneity of the melt and for reproducible closing of the non-return valve. Furthermore, the back pressure creates a constant resistance to the melt during the plasticizing process and thus generates the axial screw movement to the rear during plasticizing. During the plasticizing process under back pressure, the air trapped in the plastic pellets can escape in the direction of the feed zone and further via the hopper.

Technical literature and, in practice, machine and raw material manufacturers make a large number of recommendations concerning the optimum screw back pressure. This is because machine makers use different screw geometries with different compression ratios. Depending on the structure or division of the feed, compression and discharge zones of the screw, you need a lower back pressure in the case of a higher compression ratio and a higher back pressure in the case of a lower compression ratio. There are also differences in the type of screw to be used, such as a three-zone or a barrier screw. As a result, the specific screw back pressure will be discussed later on.

Before I go into the effects of the back pressure and the setting strategy in more detail, I would like to briefly explain the function of the screw back pressure in the following "Background knowledge" information box.

BACKGROUND KNOWLEDGE:

In order to better understand the function of the screw back pressure, the following questions must be clarified:

- Where is the screw back pressure generated?
- Where does the screw back pressure act?
- What does the screw back pressure do?
- What are the hydraulic and the specific screw back pressures?
- The screw back pressure (see Figure 5.8) is generated by the hydraulic drive cylinder of the injection molding machine and counteracts the set hydraulic pressure during the plasticizing process of the screw. The specific back pressure builds up promptly in the screw antechamber. As soon as the specific back pressure in front of the screw is greater than the counteracting hydraulic pressure, the screw plasticizes by rotating backward.

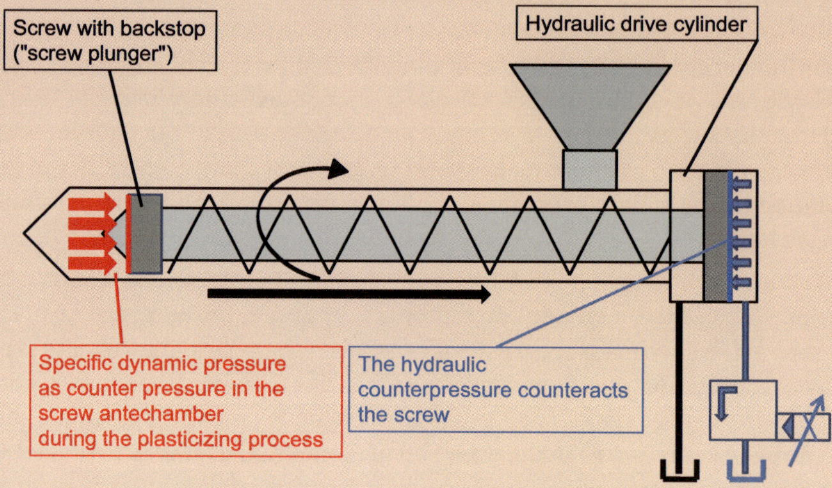

Figure 5.8 Schematic diagram of the screw back pressure

In hydraulic injection molding machines, the specific back pressure ensures that the screw can plasticize axially to the rear at the set screw circumferential speed.

In electric injection molding machines, the screw back pressure is controlled via the torque of the motor on the screw. Therefore, care should generally be taken to specify the screw back pressure as a specific pressure.

The specific screw back pressure (bar) depends on the ratio of the piston area (mm^2) of the hydraulic drive cylinder of the injection molding machine to the piston area of the screw (mm^2). As older series of injection molding machines are very often still in use, the hydraulic pressure to be set on the machine must often be determined via the required specific pressure. This can be done quickly and easily from conversion tables (Figure 5.9) usually located in the area of the plasticizing unit.

It is also possible to calculate the required hydraulic pressure from Equation 5.2:

$$\text{Hydr.pressure} = \frac{\text{Spec.dynamic pressure (bar)} \times \text{Piston area of the screw} \left(\text{cm}^2\right)}{\text{Hydraulic piston area of the machine} \left(\text{cm}^2\right)} \tag{5.2}$$

I will show you how to calculate the hydraulic back pressure using the following practical example.

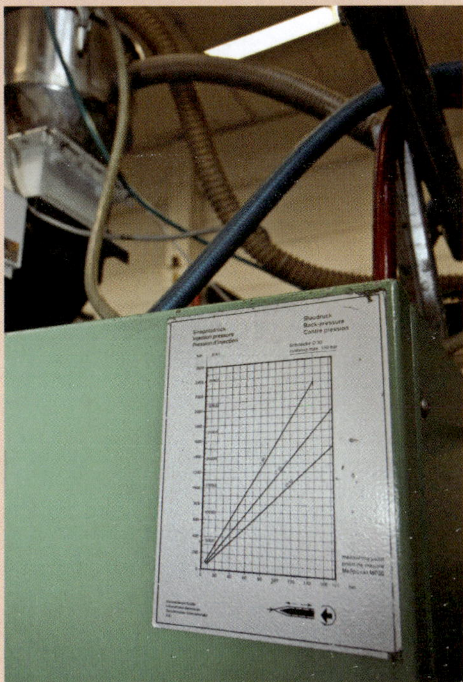

Figure 5.9 Typical conversion table for screw back pressure on the machine

PRACTICAL EXAMPLE:

Given:

- Screw diameter 50 mm → Screw piston area 19.63 cm^2
- Hydraulic piston area 313.17 cm^2 (information supplied by the machine manufacturer)
- Required specific back pressure for PC/ABS at approx. 100 bar

Sought: Hydraulic back pressure to be set in bar

Solution:

Substituting into Equation 5.3:

$$\text{Hydraulic pressure} = \frac{100\ (\text{bar}) \times 19.63\ (\text{cm}^2)}{313.17\ (\text{cm}^2)} = 6.26\ \text{bar} \tag{5.3}$$

Result:

A hydraulic back pressure of approx. 6bar must be set.

On all newer injection molding machines with microprocessor control and servo valve technology, setting the screw back pressure is very simple, as the machine automatically converts the required hydraulic pressure after the specific pressure has been entered and passes this on to the machine hydraulics via the control system.

5.2.3.1 Effects of the Screw Back Pressure

Table 5.13 summarizes the possible effects of screw back pressure on the plastic melt and the injection molding process.

Table 5.13 Effects of the Screw Back Pressure

Screw back pressure set too high:
An increase in the melt temperature via increased friction
Risk of thermal and mechanical degradation of the plastic melt
A lower plasticizing capacity and consequently a longer plasticizing time
In the case of reinforced plastics, damage to the fibers (fiber breakage)
Variations in plasticizing time
Difficulties in feeding the plastic pellets
Screw back pressure set too low:
Inhomogeneity of the plastic melt
Unmelted material
Air streaks on the molded part
Weight variations in the molded parts
Variations in the residual melt cushion

5.2.3.2 Procedure for Basic Setting of the Screw Back Pressure

Recommended screw back pressures can usually be found in the processing instructions or processing data sheets of the raw material manufacturer. If no recommended pressures are available, you can ask the raw material manufacturer's application technology department for them. The following basic instructions apply when setting the specific back pressure:

NOTE:

- Easy-flowing and easy-melting thermoplastics generally require a lower specific back pressure than difficult-flowing thermoplastics.
- When processing fiber-reinforced thermoplastics, particularly low specific back pressures should be set to prevent mechanical damage to the fibers. There is a high risk of the fibers breaking and shortening. As a result, a molded part, for example, will have lower mechanical strength and greater distortion.

To help you determine the required specific back pressure, a possible procedure is shown below:

- **Step 1:**

  The recommended specific screw back pressures are generally given in a defined value range, e. g. 50–100 bar for polystyrene (PS). You should always start with the lowest recommended specific back pressure

- **Step 2:**

  Start plasticizing at the lowest recommended value and record the required plasticizing time in writing

- **Step 3:**

- Repeat this process several times by plasticizing the material and injecting it again. Any fluctuations in the plasticizing time will now become apparent. If this is the case, increase the specific back pressure by a further 40–50%. It is important to note that the plasticizing time should only be slightly longer, by 5–15%, when increasing the back pressure

- **Step 4:**

  If the plasticizing time is constant and slightly longer, the correct specific screw back pressure has been set

- **Step 5:**

  The final step is to evaluate the plastic melt after injecting (see Section 5.7 "Evaluating the plastic melt")

You will see that, after determining the specific screw back pressure, the values will usually be around 100 bar. This can also be taken as a guide for common thermoplastics. However, a recommended value from the raw material manufacturer should always be known and you should not exceed or fall below such values.

5.2.4 Setting the Screw Decompression

The last parameter to be set for the plasticizing process is screw decompression, also known as screw retraction. Screw decompression reduces the pressure in the screw antechamber and supports the mechanical closing of the non-return valve. During the injection process, the non-return valve prevents the melt from flowing back into the screw flights.

If a mold with a hot runner system (HRS) is to be sampled, you should also set a screw decompression with open nozzles before the plasticizing process. This ensures pressure relief in the hot runner system. If material runs out of the nozzle during the pressure relief of the hot runner system, the screw retraction must be increased.

Unfortunately, it is often the case in injection molding shops that the screw retraction is set too high or not at all. Although excessive screw retraction reduces the pressure in the screw antechamber, it also ensures that air enters the front area of the barrel. This is usually characterized by black dots (soot particles) or streaks on the part, as the air in the front area of the barrel is compressed and heated so much during the injection process that the diesel effect occurs and the melt burns.

Unfortunately, there is no exact specified value for screw retraction; a great deal of experience and prevailing conditions, such as material behavior, play an important role here. In practice, a screw retraction of approx. 10mm at a retraction speed of 10mm/s has proven to be a good guideline.

PRACTICAL TIP:

Many specialists immediately set a larger screw retraction if, for example, melt runs out of the nozzle after screw decompression. In such cases, do not immediately increase the retraction, but instead check whether the material has been properly pre-dried as per the material data sheet. The set nozzle temperature should also be checked for correctness and finally a functional test of the nozzle heating should be carried out.

5.3 Correctly Setting the Injection Phase

The injection process is a decisive process step, because it has a major influence on the quality characteristics of the molded part, such as surface quality, orientation in the surface layer and crystallinity.

The injection process is made up of the following parameters:

- Required injection pressure (bar)
- Injection time (s) as a function of the required injection speed (mm/s) or the required injection volume flow rate (cm^3/s)
- Injection profiling adapted to the molded part

IMPORTANT NOTE:

Depending on the machine setup, the injection speed in mm/s or the injection volume flow rate in cm^3/s may be used.

In the rest of this chapter, we will mainly work with the injection volume flow rate (cm^3/s).

The following subsections explain the necessary basic knowledge, the influences and effects as well as a recommended procedure for these very important process parameters in more detail.

5.3.1 Injection Pressure

The injection pressure (bar) is the highest pressure exerted on the plastic melt during injection via the screw plunger. As there is a pressure ratio of the hydraulic drive piston of the machine to the piston surface of the plasticizing screw analogous to the screw back pressure, the specific injection pressure prevailing in the screw antechamber is also used here.

The specific injection pressure is directly related to the injection volume flow rate (cm^3/s). If a high volume flow rate is set for injection, the machine always requires a higher pressure to realize this. Additional factors such as the flow behavior of the plastic, the mold and melt temperature and the flow path/wall thickness ratio also play a decisive role in the required pressure level. Generally speaking, the higher the flow resistance during filling, the higher the required injection pressure (bar).

In modern injection molding machines, the machine automatically sets the specific injection pressure (bar) required to achieve the set injection volume flow rate (cm^3/s). In order to ensure that the machine maintains the set injection volume flow rate (cm^3/s) at a constant level at the start of sampling, the maximum specific injection pressure (bar) should first be set on the machine as the basic setting.

NOTE:
Do not forget to limit the injection pressure after completing the filling study (Chapter 6)!

To protect the mold, it is advisable to limit the required specific injection pressure (target value) again after the injection process has been optimized. This limitation of the injection pressure (bar) serves to avoid the risk of over-injection or overloading of the molded part if, for example, the plasticizing volume is set too high. The pressure should be 10–20% above the required injection pressure (bar) in order to be able to compensate for possible viscosity fluctuations in the melt.

5.3.2 Injection Profile

The injection profile should be individually adjusted in several stages to suit the molded part. The aim must be to ensure that the flow front speed remains as constant as possible as a function of volume flow rate. As we are usually dealing with complex molded parts of different wall thicknesses, it is particularly important to adapt the profile to these conditions, e. g. by reducing the injection volume flow rate (cm^3/s) before the flow front reaches the wall thickness difference (Figure 5.10). These adjustments to the wall thickness differences of the molded part ensure that filling is as constant as possible.

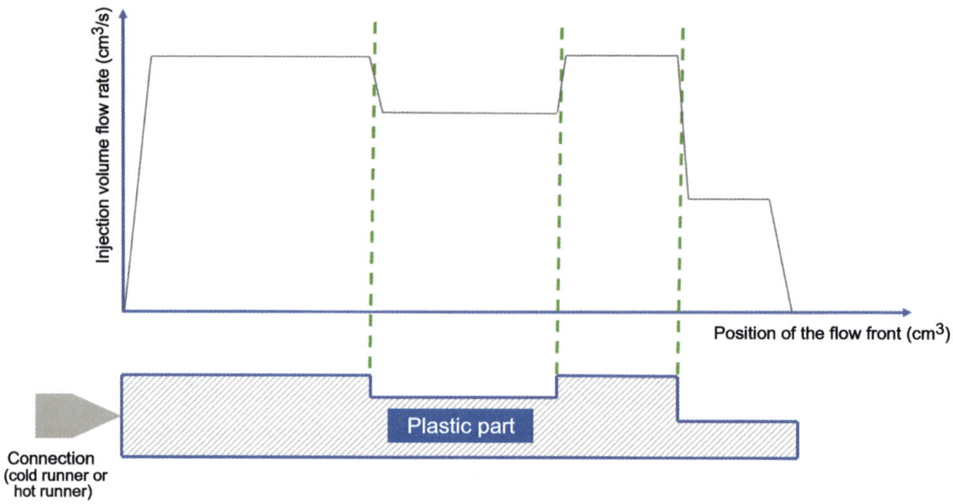

Figure 5.10 Adapting the injection profile to the molded part

Plastic injection-molded parts should always be designed to be compatible with the plastic, by designing the molded part to have the same wall thicknesses. Unfortunately, this is usually just theory and cannot be realized in practice. As a result, the molded part may have extremely thin areas. Where very thin-walled areas are present, a high volume flow rate must be used. This has the advantage that heat is generated which can compensate for heat dissipation to the mold wall.

You can now see that the injection profile depends on a number of important factors. Basically, a slow—fast—slow profile has proved very successful. Such a three-stage profile (slow—fast—slow) basic setting is explained in more detail below (see also Figure 5.11).

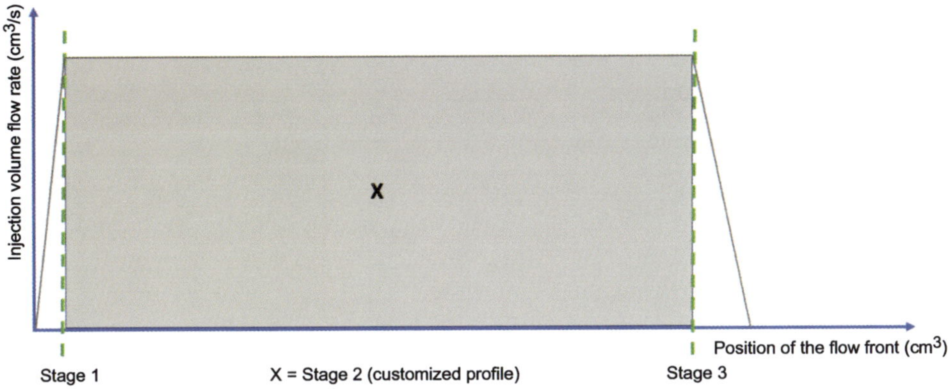

Figure 5.11 Three-stage injection profile as basic setting

Stage 1: A slower volume flow rate at the start of the injection phase:

- Reduces deformation of the mold
- Lowers the risk of jetting
- Reduces the tendency for matt or cloud-like areas in the gate area of the molded part

Stage 2: Volume flow rate adapted to the cross-section between stage 1 and stage 3:

- Additional stages may be installed in stage 2, depending on the injection molded part

Stage 3: A slower volume flow rate at the end of the injection phase:

- Improves the switching accuracy from injection to holding pressure and thus the part quality
- Enables a reduction in clamping force
- Improves the venting properties of the mold (avoidance of diesel effects)
- Dampens or absorbs the dynamic movement of the plasticizing screw
- Avoids pressure peaks, and
- Is gentle on the mold

In order to be able to determine an optimum profile for your molded part, a filling study of the molded part is required. This allows you to identify all relevant molded part areas on the basis of partial fillings and thus assign the required profile. The filling study is dealt with in detail in Chapter 6 "Filling study, holding pressure and mold clamping force" of the mold sampling phase.

5.3.3 Injection Volume Flow Rate (cm^3/s)

Goal: To inject the plastic into the mold at the correct volume flow rate (cm^3/s).

In many companies, this process step is currently tried out on the machine (by trial and error) rather than being professionally analyzed and optimally adjusted in line with standardized criteria and later optimized.

The major problem with this is that the development of the injection phase (optimization routine) is based on the wealth of experience accumulated by the respective employee.

Problem: Experience = Sum of all faults

In addition to subsequent process fluctuations, this results in a considerable increase in use of resources, as more time is required to find an optimally functioning injection phase specific to the case at hand. Furthermore, increased costs are incurred, due to the large number of rejects produced by *"trial and error"*.

The injection volume flow rate (cm^3/s) indicates how much volume of plastic is transported through a defined cross-section of the molded part per unit time. It also influences a number of quality characteristics of the molded part. For example, a high volume flow rate leads to molecular orientations. The reason is that, before injection begins, the melt is initially present as disordered molecules, which then stretch and orientate themselves in the direction of flow during the injection process. These orientations lead to anisotropy (non-uniform properties in the molded part, e.g. uneven shrinkage) and to a higher warpage potential. Furthermore, the volume flow rate may, to an extent depending on the quality requirements, adversely affect the surface properties, such as the visibility of filling and weld lines, differences in gloss etc.

The injection volume flow rate (cm^3/s) also influences the flow behavior. For example, if a very high flow rate is used for filling, the flow front behavior will be different from that of a low flow rate. It is therefore also possible to influence the position of filling or weld lines and their properties with the help of the resulting volume flow rate. The molded part dimensions and the molded part weight can also be influenced by the injection volume flow rate (cm^3/s).

Nowadays, simulations can be used to study the influence of the volume flow rate on properties such as orientation effects, shear stresses, and shear rate in the molded part. Unfortunately, however, the influence of the volume flow rate on the optical characteristics, such as gloss, surface quality, and streaks, on the molded part cannot be described mathematically.

5.3.3.1 Effects of the Injection Phase

To give you a better feel for the injection phase, the possible influences and effects of too low or too high a volume flow rate on the molded part and the melt are summarized in Table 5.14:

Table 5.14 Effects of the Injection Phase

Injection phase too slow:	
Effect:	**Consequence:**
The molded part is not fully injected	The melt flows too slowly into thin-walled areas or areas far from the gate. The flow front comes to a standstill and solidifies (frozen melt).
Visibility of flow and weld lines	Optical notches and consequently loss of strength occur due to cooled edge layers of the melt.
Cold displacement	The already solidified melt on the mold wall is displaced during further filling, resulting in transverse grooves perpendicular to the direction of flow (vinyl record effect).
Variations in the weight of the molded part	Irregular closing of the non-return valve, as the locking ring does not assume the required position during injection.
Greater cooling of the melt	Higher temperature differences between the gate area and the end of the flow path.
Low melt flow velocity	Less molecular alignment; less frictional heat.
Injection phase too fast:	
Effect:	**Consequence:**
The air cannot escape from the mold fast enough	The air is highly compressed and burns due to the high temperature (diesel effect).
Jetting, characterized by a visible "worm-like" appearance from the gate area into the molded part	The melt is injected freely into the cavity without undergoing fountain flow.
Formation of streaks	Air inclusions on the surface of the molded part, due to non-uniform flow front.
A higher clamping force is required	Increased mold breathing; risk of flash formation in the mold.

Table 5.14 Effects of the Injection Phase (*continued*)

High shear stresses occur	Surface defects in the gate area (usually "cloud-like").
High melt flow velocity	Increasing molecular orientation (anisotropy in the molded part, resulting in a higher tendency to warp).
A high velocity gradient between the mold wall and the plastic core of the melt	Shear stress or shear fracture, which can damage the plastic.
High shear stress	Damage to the plastic; leads to increased fiber breakage in fiber-reinforced materials.

5.3.3.2 Determining the Required Injection Volume Flow Rate (cm^3/s)

Before we go into possible solutions for determining the required injection volume flow rate (cm^3/s), I would like to briefly introduce you to the world of rheology with the help of the following background knowledge.

BACKGROUND KNOWLEDGE:

What is rheology?

The word rheology is derived from the Greek word "*rheos*" = to flow. Rheology is therefore the science of the deformation and flow of materials.

Flow can be understood as the continuous deformation of a material under the influence of external forces. The task of rheology is to describe, measure and explain how a solid or a liquid reacts to a certain deformation.

What is viscosity?

Viscosity is a measure of the plastic's resistance to flow. The higher the viscosity, the thicker (less flowable) the plastic is; the lower the viscosity, the thinner (more flowable) it is.

What is meant by shear rate?

Shear stress is a force that acts over a given area of the fluid and gives rise to a velocity gradient across the thickness (e. g. plastic melt in the cavity) known as the shear speed or shear rate. The velocity gradient indicates how fast the overlapping intermediate layers (e. g. the plastic melt during the filling process) move relative to each other. This ratio describes the rate at which the plastic undergoes shear, for example; hence the term "shear rate".

Plastics are viscoelastic materials!

The viscosity of plastic melts decreases as the shear rate increases (Figure 5.12). The melt becomes "thinner" because the structure of the molecules changes as the shear rate increases during flow.

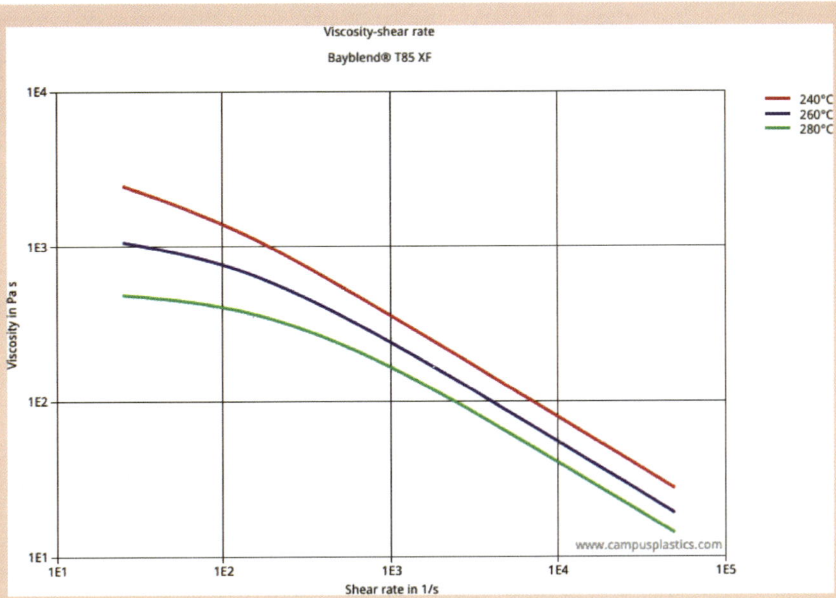

Figure 5.12 Example of a viscosity curve for PC/ABS

The tangle of molecules becomes increasingly less tangled and the parallel molecules are able to slide off each other more easily. If thermal energy is applied in the form of melting, it first forces its way between the threads of molecules and thus weakens the electrostatically induced secondary valence forces. As it is the ordered structure of the semi-crystalline materials that initially breaks down, all melted thermoplastics are amorphous, a fact which is important for determining the volume flow rate for filling the cavity later on.

This behavior is referred to as *"structural viscosity"*. The structural viscosity is an important and essential prerequisite for injection molding. It is the only reason why it is possible to inject large quantities of molten plastic through the smallest channels and cavities, for example, at realizable injection pressures. Recall that the gate diameters of a cold runner can sometimes be as small as 0.5–1.0mm [1].

Today's literature or training courses do not have a straightforward and highly effective way of determining the optimum injection phase. Good rheology literature resorts to higher mathematics and physics, but these cannot be performed during sampling on the machine.

However, there are a good many solutions that utilize injection speeds (mm/s) and injection volume flow rate (cm^3/s). We present an approach below which utilizes different methods that take account of the rheological and thermal effects on one hand and, on the other, are comprehensible in practice and, above all, can be performed quickly and easily by the specialist.

In practice, many machine manufacturers have established a maximum injection speed of approx. 100 mm/s (i. e. 1 s filling time per 100 mm) for standard machines, for example.

The following criteria can be used to decide:

- Slowly up to 30% (thick-walled parts, e. g. larger than 4 mm)
- Moderately from 30–70% (technical parts)
- Quickly above 70% for thin-walled parts (e. g. 1–1.5 mm)
- Packaging: from 300 mm/s upward
- The whole thing then needs to be refined in respect of ribs, thin points, part requirements, etc.
- You then arrive at the filling time via the shot volume and the selected screw diameter.

In the following section, we will return to a selected solution approach for determining the required injection volume flow (cm^3/s).

5.3.3.2.1 Suggested Approach: Injection Time (s) as a Function of Injection Volume Rate (cm^3) and Viscosity of the Plastic

- **Step 1:**

  Determine the required injection volume (cm^3) of the molded part → Take the output factor into account!

- **Step 2:**

  Determine the injection time (s)

  Method 1: *Determine the injection time using computer simulation results*

  Today's computer-aided simulation programs can be used to calculate flow processes under the corresponding boundary conditions. However, it is important here that all relevant data for the setup, such as rheological plastic values, be available and correctly stored in order that realistic values may be obtained.

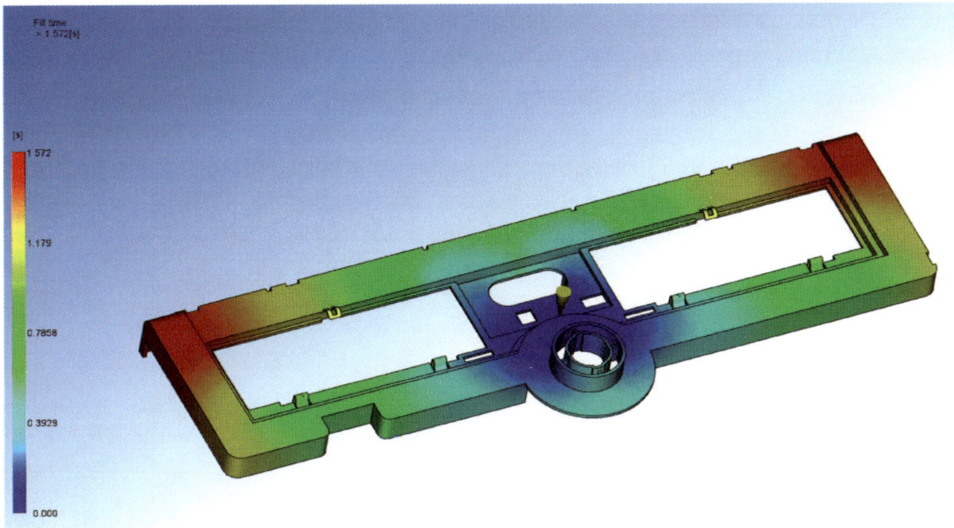

Figure 5.13 Determining the filling time via a computer simulation

Method 2: *Take the injection time (s) as a function of the injection volume (cm^3) and viscosity of the plastic from* Table 5.15

Table 5.15 Example of Standard Values for the Injection Time (*as a function of injection volume (cm^3) and viscosity of the plastic*)

Injection volume (cm^3)	Injection time (s)		
	PA 4.6, PA 6, PA 66, PA 6.10, PA 11, POM, PET, PBT, PPS, TPE, PE plasticized	PS, SB, SAN, PP, ABS, PA 12, PA amorphous, PPE, CA, CAB, CP, PE hard	PC, PMMA, PAI, PEI, PVC hart, PSO, PES, FEP, ETFE
1–8	0.10–0.25	0.25–0.35	0.30–0.50
8–15	0.25–0.35	0.35–0.55	0.55–0.70
15–30	0.35–0.50	0.55–0.70	0.70–0.90
30–50	0.50–0.65	0.70–0.90	0.90–1.20
50–80	0.65–1.00	0.90–1.30	1.20–1.80
80–120	1.00–1.60	1.30–2.00	1.80–2.80
120–180	1.60–2.50	2.00–3.00	2.80–4.20
180–250	2.50–3.50	3.00–4.20	4.20–5.70
250–350	3.50–4.60	4.10–6.60	5.70–7.50

Table 5.15 Example of Standard Values for the Injection Time (*as a function of injection volume (cm³) and viscosity of the plastic*) (*continued*)

Injection volume (cm³)	Injection time (s)		
	PA 4.6, PA 6, PA 66, PA 6.10, PA 11, POM, PET, PBT, PPS, TPE, PE plasticized	PS, SB, SAN, PP, ABS, PA 12, PA amorphous, PPE, CA, CAB, CP, PE hard	PC, PMMA, PAI, PEI, PVC hart, PSO, PES, FEP, ETFE
350–550	4.60–6.50	6.60–8.00	7.50–9.50
550–850	6.50–9.50	8.00–10.50	9.50–12.50
850–1200	9.50–12.50	10.50–13.50	12.50–15.50
1200–1650	12.50–16.00	13.50–17.00	15.50–19.50
1650–2200	16.00–20.00	17.00–21.00	19.50–23.50
2200–3000	20. 00–24.00	21.00–25.50	23.50–28.00

PRACTICAL EXAMPLE:

Method 2: *Guide values for the injection time (s) as a function of injection volume (cm³) and viscosity of the plastic from* Table 5.15

Given:

- Housing cover made of ABS
- Molded part weight: 183 g → Calculated volume of the molded part taking into account the output factor (see Section 5.2) at 198.91 cm³

Sought: Injection volume flow rate (cm³/s)

Solution:

Step 1: Take the standard value for injection time (s) from Table 5.15 for ABS at 198.91 cm³

→ Injection time selected at 3.5 s

Step 2: Calculate the required injection volume rate (cm³)

$$\dot{V} = \frac{\text{Required injection volume } (\text{cm}^3)}{\text{Injection time (s)}} = \frac{198.91(\text{cm}^3)}{3.5(\text{s})}$$
$$= 56.83\frac{\text{cm}^3}{\text{s}} \approx 57\frac{\text{cm}^3}{\text{s}}$$

Method 3: *Calculation of the required injection time (s) via dynamic filling*

To calculate the dynamic filling, we need the "minimum wall thickness" of the molded part. With its help, we can calculate the time at which the mold must be filled volumetrically.

Each plastic has a specific effective thermal diffusivity (α_{eff}) whose units are mm²/s. The effective thermal diffusivity is the quotient of the thermal conductivity and the product of the density times the specific heat capacity. Let's take a look at two different thermoplastics. Firstly, polycarbonate, which has a density of 1.21 g/ cm^3 at room temperature. If it is now melted at 300 °C, the density is only 0.95 g/ cm^3. If PC is injected into a mold with a mold wall temperature of 60 °C, α_{eff} is 0.1087 mm^2/s. For a mold with a temperature of 100 °C, α_{eff} is 0.115 mm^2/s. The effective thermal diffusivity therefore barely changes. The situation is somewhat different for PE-HD. In a mold at 10 °C, α_{eff} is 0.1015 mm^2/s and at a mold temperature of 60 °C α_{eff} is only 0.0684 mm^2/s. This represents a change of 32.6%. As α_{eff} is part of a factor later in the calculation, it is advisable to create a table of values for the plastics you are working with so that there are no significant deviations between the calculation and reality later on. Now let's look at the filling process. If you look at the injection curve, you will see a small deflection in the starting area, which then drops off again. At this point, the PSE's locking ring closes. Up to this point, nothing has happened in the melt or during the filling process. This is followed by the so-called dynamic filling phase. There is now a cavity in the mold and the melted plastic (which now has the output factor as its density) fills this cavity until the mold is completely filled volumetrically. It is important that the plastic fills all areas with a reasonably equal viscosity, e. g. to avoid stresses. To calculate the dynamic filling, we need the smallest wall thickness of the part.

Since every thermoplastic is amorphous in the molten state, we use Equation 5.4:

$$t = \frac{s^2}{\pi^2 \times \alpha_{eff}} \times \ln\left[\frac{4}{\pi} \times \left(\frac{T_{Melt} - T_{Mold}}{T_D - T_{Mold}}\right)\right] \tag{5.4}$$

s = Minimum wall thickness

α_{eff} = Effective thermal diffusivity of the material (mm^2/s)

T_{Melt} = Temperature of the plastic melt (°C)

T_{Mold} = Temperature of the mold (°C)

T_D = Demolding temperature of the molded part (°C)

Here we use the "*minimum wall thickness*" of the molded part. This makes it easy to calculate the duration of the dynamic filling. Each corner of the injection molded part is now filled with plastic of almost identical viscosity.

Consequently, all areas of the molded part walls are frozen to such an extent that the plastic core, the plastic of which has to be compressed to its cold density, is still present. This wall thickness, which is now still open in the form of the plastic core, is subsequently compressed during the holding pressure phase (more about this in Chapter 6) and cooled over the required residual cooling time.

If the thickness of the current plastic core is known, the holding pressure phase could also be calculated further by the same approach [2].

- **Step 3:**

  Calculate the required injection volume flow rate (cm³/s)

$$\dot{V} = \frac{\text{Injection volume } (\text{cm}^3)}{\text{Determined injection time } (\text{s})} \tag{5.5}$$

- **Step 4:**

  Set the calculated injection volume flow rate (cm³/s) in the injection molding machine with adapted basic injection profile

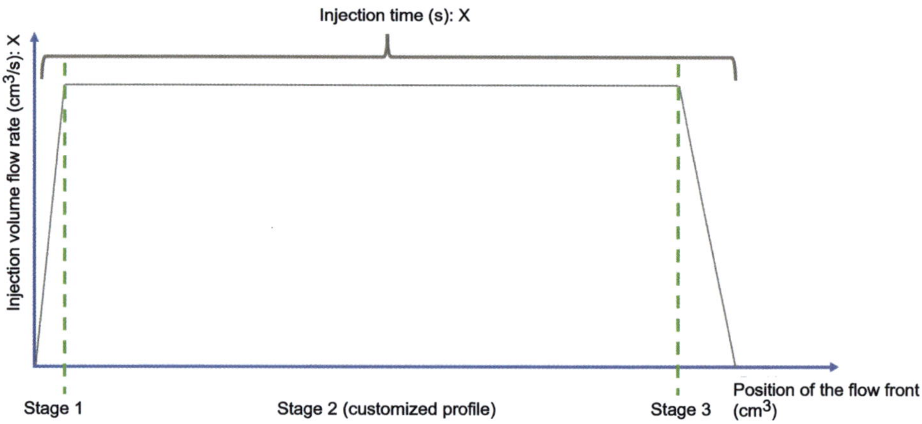

Figure 5.14 Sample injection basic setting using the determined values x

- **Step 5:**

  In the subsequent filling study (more about this in Chapter 6), the basic injection profile (Figure 5.14) in the area of stage 2 will be individually adapted to the molded part

5.3.4 Injection Time (s)

The resulting injection time (s) results from the set volume flow rate (cm³/s) and the required flow path of the melt and should be used as a monitoring parameter in the injection molding machine. Injection time monitoring interrupts the process if the set time is exceeded and so the machine switches to fault mode. This monitoring basically serves to ensure that the mold cannot be over-injected, for example if the plasticizing volume selected is too large.

The injection time can draw attention to other sources of error in the injection molding process, such as changes in the melt and mold temperatures during the process.

Changes caused by incorrect settings change the flow resistance of the melt and therefore also the required injection time. However, it is also possible to draw conclusions about the constant plastic quality, as the melt index also changes when the general plastic properties change, resulting in a different flow property.

5.4 Cooling Time and Demolding Temperature

The following sections describe a structured approach to achieve the optimum basic setting for the required "cooling time", which must be set on the machine.

5.4.1 Cooling Time

The cooling time (s) begins as soon as the hot plastic melt touches the colder mold wall in the cavity during the injection process. The cooling time is divided into two phases (Figure 5.15), the holding pressure time and the time required afterward for the demolding temperature to be reached. The latter is known as the residual cooling time and is set on the machine.

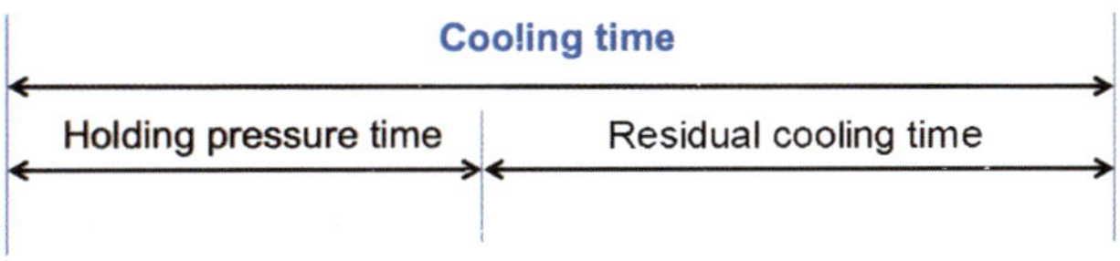

Figure 5.15 Dividing the cooling time into two phases

The main factors influencing the cooling time are the wall thickness (mm) of the molded part, the temperature conductivity (mm^2/s) of the plastic, the set material and mold temperature (°C) and the demolding temperature (°C).

Unfortunately, in practice, processes are often run with excessively long residual cooling times, even though the demolding temperature of the plastic has long since been exceeded. One reason for this is that a longer residual cooling time was selected in the sampling phase for safety reasons and no attention was paid to this in the further course of the sampling process. Excessively long residual cooling times have a negative effect on the economics and energy efficiency of the process, as this results in an unnecessary extension of the cycle time and higher specific energy consumption. Taking into account the required part quality, the shortest required residual cooling time must therefore always be aimed for. If the residual cooling time is set too short, ejector marks, deformation during ejection and distortion of the molded part will occur.

5.4.2 Demolding Temperature of the Plastic

The demolding temperature (T_D) is the temperature that prevails in the middle layer of the wall thickness of the molded part at the time of demolding. The demolding temperature also describes the temperature at which the plastic is sufficiently solid to be demolded without permanent deformation or distortion.

To be able to set an optimum residual cooling time for your molded part, you need the demolding temperature of the plastic to be processed. This immediately raises the question: "Where do you get a practical demolding temperature?"

Firstly, there is the material manufacturer. The material manufacturer usually specifies a recommended average demolding temperature in its processing data sheets. If there are no recommended values on the data sheet, you can ask your raw material manufacturer's application technology department for them.

You can also use the plastic's shear modulus temperature curve to determine a standard value. The molded part may only be demolded if the average demolding temperature is in a range in which the plastic can no longer "flow" or absorb forces. Consequently, this temperature can be read from the shear modulus curve of the plastic. Figure 5.16 shows the shear modulus curve for the example of PC/ABS. Here you can see that above a temperature of approx. 110 °C the shear modulus drops steeply. The point at which the drop occurs indicates that dimensional stability is no longer guaranteed. Consequently, the maximum demolding temperature that the molded part must reach is approx. 110 °C.

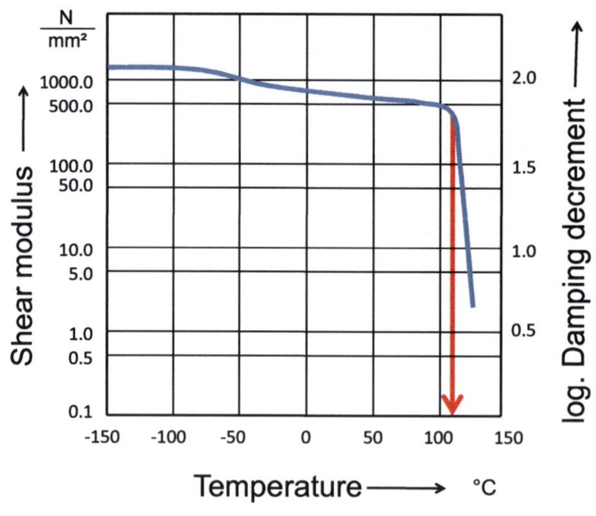

Figure 5.16 Determination of the demolding temperature from a shear modulus temperature curve for the example of PC/ABS

As a last resort, you can use the Vicat softening temperature (VST) as a standard value for the demolding temperature. The VST can usually be found under thermal properties in the technical material data sheet of the raw material manufacturer (Figure 5.17). For the PC/ABS example, the Vicat softening temperature according to VST/A/50 ISO 306 is approx. 128 °C.

Thermal properties

C	Temperature of deflection under load	1.80 MPa	°C	ISO 75-1,-2	107
C	Temperature of deflection under load	0.45 MPa	°C	ISO 75-1,-2	126
C	Vicat softening temperature	50 N; 50 °C/h	°C	ISO 306	128
	Vicat softening temperature	50 N; 120 °C/h	°C	ISO 306	130
C	Coefficient of linear thermal expansion, parallel	23 to 55 °C	10^{-4}/K	ISO 11359-1,-2	0.75
C	Coefficient of linear thermal expansion, normal	23 to 55 °C	10^{-4}/K	ISO 11359-1,-2	0.8
C	Burning behavior UL 94 [UL recognition]	0.85 mm	Class	UL 94	HB
C	Oxygen index	Method A	%	ISO 4589-2	24

Figure 5.17 Excerpt from a technical material data sheet from a raw material manufacturer (image source: Covestro)

5.4.3 Basic Setting of the Cooling Time

Once you have determined the required demolding temperature for the plastic to be processed, the next step is to set the basic cooling time. The following methods are described in more detail:

- **Method 1:** Cooling time determined via computer simulation
- **Method 2:** Estimation using an approximation formula

5.4.3.1 Cooling Time Determined Via Computer Simulation

Nowadays, the cooling process in a three-dimensional molded part can be precisely estimated using modern injection molding simulation (Figure 5.18). If your molded part was simulated in the development or design phase, you can extract important information from this for your sampling, such as the time required to reach the demolding temperature and the temperature distribution in the molded part. In our example, the time required to reach the demolding temperature is approx. 25 °C.

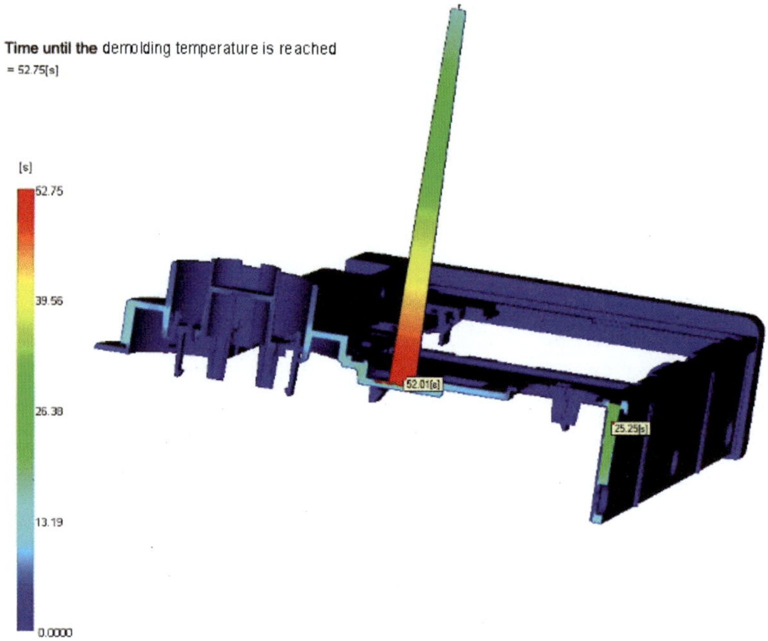

Figure 5.18 Computer simulation of cooling time

5.4.3.2 Estimation Using an Approximation Equation

If you do not have results from an injection molding simulation, an approximation equation can be used to estimate the cooling time. Equation 5.6 was developed for one-dimensional geometries and makes it possible to calculate cooling processes in a specific spatial direction, e.g. the molded part thickness, until the average temperature has reached the demolding temperature.

$$t_C = \frac{s^2}{\pi^2 \times \alpha_{\text{eff}}} \times \ln\left[\frac{8}{\pi} \times \left(\frac{T_{\text{Melt}} - T_{\text{Mold}}}{T_D - T_{\text{Mold}}}\right)\right] \tag{5.6}$$

Where:

t_C = Cooling time (s)

s = Maximum wall thickness of the molded part (mm)

α_{eff} = Effective thermal conductivity (mm^2/s) (see Table 5.16) is made up of thermal conductivity, density and specific heat capacity

T_{Melt} = Melt temperature (°C). Use manually measured melt temperature

T_{Mold} = Average mold temperature (°C) Use the manually measured mold wall temperature

T_D = Average demolding temperature. Use the required demolding temperature (see Section 5.4.2 "Demolding temperature")

Table 5.16 Effective Thermal Conductivity Coefficients for Thermoplastics

Thermoplastic	α_{eff} (mm²/s)	Thermoplastic	α_{eff} (mm²/s)
ABS	0.07	PET	0.09
PC/ABS	0.09	PMMA	0.08
PA amorphous	0.07	POM	0.05
PA 6	0.07	PP	0.06
PA 6.6	0.08	PP with glass fiber content	0.09
PA 6 with glass fiber content	0.09	PS	0.09
PA 6.6 with glass fiber content	0.12	SAN	0.08
PBT	0.08	TPE	0.08
PE-HD	0.06	TPU	0.08
PE-LD	0.05	PC	0.11
ASA	0.08		

The approximation equation (Equation 5.6) utilizes average temperatures. These are the mean values of all temperatures over the cross-section of the molded part or mold. Unfortunately, it is not possible for the "mold sampling specialist" to determine the exact average temperatures for the molded part and the mold. There is also a lack of precise knowledge about the cooling efficiency of the mold and the different temperature distribution in the molded part between the gate and the end of the flow path. As this is an approximation equation and the result is only supposed to serve as an initial value for the cooling time, you can take the demolding temperature (Section 5.4.2 "Demolding temperature of the plastic") of your plastic as the average demolding temperature. Take the set mold temperature as the average mold temperature. The melt temperature is obtained by manually measuring the molded plastic melt.

The "Practical example" box below shows the calculation process step by step for the PC/ABS housing cover.

PRACTICAL EXAMPLE:

Given:

- Housing cover made of PC/ABS
- $s = 2.3\,\text{mm}$
- $T_{Mold} = 70\,°\text{C}$
- $T_{Melt} = 260\,°\text{C}$
- a_{eff} = forPC/ABS in Table 5.16 → $0.09\,\text{mm}^2/\text{s}$
- $T_D = 105\,°\text{C}$

Sought: T_C (s)

Solution:

Substituting into Equation 5.6:

$$t_C = \frac{2.3^2}{\pi^2 \times 0.09\frac{\text{mm}^2}{\text{s}}} \times \ln\left[\frac{8}{\pi} \times \left(\frac{260°\text{C} - 70°\text{C}}{105°\text{C} - 70°\text{C}}\right)\right] = 15.67\ \text{s}$$

Answer:

The basic setting is a cooling time of approx. 15.67 s on the machine.

The approximation equation for determining the cooling time (Equation 5.6) shows that the greatest influence on the cooling time is exerted by the mold temperature, the material-dependent demolding temperature and the wall thickness. The cooling time increases by the square of the wall thickness. If the wall thickness changes by a factor of 2, the cooling time increases by a factor of 4. If the mold temperature is increased by just 10 °C, this can result in an increase in cooling time of up to 15%. The same result is obtained if the demolding temperature is increased by 10 °C. Increasing the mold temperature by 10 °C has a minor effect of increasing the cooling time by up to 5%.

NOTE:

As this is an estimate of the cooling time (without holding pressure time), it is very important in the following Chapter 6 "Filling study, holding pressure and mold clamping force" of the sampling process to check the molded parts for their required demolding temperature during the filling study and to adjust the cooling time if necessary. An exact check of the cooling time is carried out using thermal imaging (thermography) in the subsequent mold sampling analysis (Chapter 7 "Mold sampling analysis of the basic setting").

5.5 Correct Nozzle Contact Point and Settings

Once all the basic settings of the plasticizing unit have been made, the plasticizing unit must be moved onto the mold in set-up mode and zeroed.

The optimum nozzle contact point must be checked by means of a nozzle impression. The goal here is to verify that the radii of the mold nozzle and machine nozzle are a functional match. Important criteria are that the radius of the nozzle is always smaller than the radius of the mold and the diameter of the nozzle bore is smaller than the diameter of the sprue bushing in the mold. The two possible fault patterns and an optimum nozzle arrangement on the mold are shown schematically below.

- If the nozzle radius is larger than that of the sprue bushing of the mold (Figure 5.19), plastic melt is deposited between the sprue bushing and the nozzle. Consequently, demolding of the sprue is not possible.

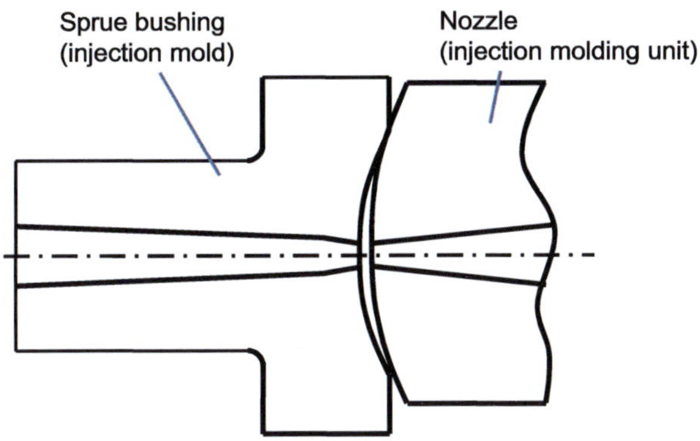

Figure 5.19 Nozzle radius larger than that of the sprue bushing of the mold

- If the diameter of the nozzle bore is larger than that of the sprue bushing of the mold (Figure 5.20), plastic melt is deposited between the sprue bushing and the nozzle. As a result, demolding of the sprue is not possible.

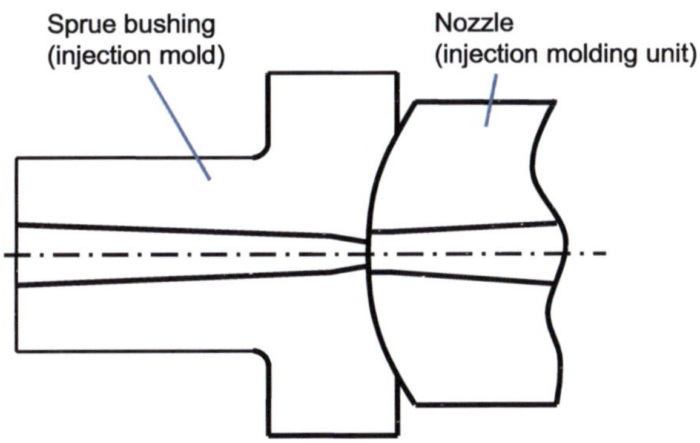

Figure 5.20 Diameter of the nozzle bore larger than that of the sprue bushing of the mold

- Figure 5.21 shows optimum nozzle contact with the mold, as the radius of the nozzle is smaller and the diameter of the nozzle bore is also smaller than the bore of the sprue bushing.

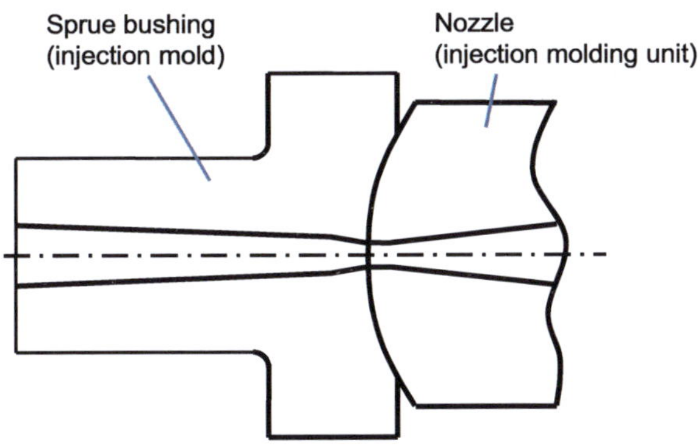

Figure 5.21 Optimal nozzle contact with the mold

5.5.1 Setting the Nozzle Contact Force

The nozzle contact force ensures that the plastic melt does not escape into the open at the contact surface between the nozzle and the sprue bushing of the injection mold. Companies often pay no attention to the nozzle contact force, which means that it is usually far too high or that the maximum nozzle contact pressure is used. These unnecessary contact forces ensure that the nozzle and the sprue bushing of the mold can be compressed or damaged.

It is not easy to make a specific recommendation for the nozzle contact force to be set, as the values given vary from one machine manufacturer to another manufacturer for the same machine size. The recommended nozzle contact forces shown in Table 5.17 are to be regarded as guide values [3].

Table 5.17 Nozzle Contact Forces of Injection Molding Machines [3]

Clamping force (kN):	Nozzle contact force (kN):
Up to 500	30 to 100
501 to 1000	50 to 100
1001 to 5000	80 to 200
5001 to 10 000	120 to 250
Over 10 000	120 to 400

5.5.2 Procedure for Creating a Nozzle Impression

- **Step 1:** Hang paper between sprue bushing and machine nozzle
- **Step 2:** Move the machine nozzle onto the mold in set-up mode
- **Step 3:** Assess nozzle centering and radii using the imprint on the paper (Figure 5.22)

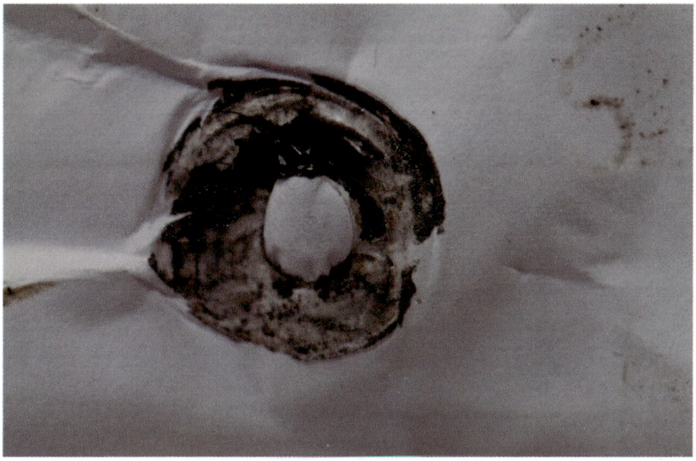

Figure 5.22 Example of a nozzle imprint

5.6 Setting the Movement of the Plasticizing Unit

In many cases, usually with cold runner molds, it is necessary to allow the plasticizing unit to "lift off" (axial movement of the plasticizing unit). This is particularly useful for thermal separation between the colder injection mold and the hot machine nozzle. For example, it can avoid freezing effects on the machine nozzle or possible stringing of the plastic. Hot runner molds should generally be operated with the machine nozzle in contact. Whether it is necessary to "lift off" the plasticizing unit must therefore be assessed at the machine by the "mold sampling specialist". The "lift-off movement" should be 10–15mm.

5.6.1 Speed of Movement of the Plasticizing Unit

If the nozzle needs to be "lifted off", this must be done at high speed, but without jolts. Care must be taken here to ensure that no tilting or other deviations from the center axis of the machine occur during acceleration and deceleration. The machine nozzle must be placed gently on the sprue bushing of the mold at slow speed, as the nozzle is easily forged shut or damaged by hard impact. The recommended average movement speeds of the plasticizing unit in Table 5.18 are guide values [3].

Table 5.18 Movement Speeds of the Plasticizing Unit [3]

Clamping force (kN):	Movement speed range (mm/s):
Up to 500	> 0 to 350
501 to 1000	> 0 to 300
1001 to 5000	> 0 to 250
5001 to 10,000	> 0 to 150
Over 10,000	> 0 to 130

5.7 Assessment of the Plastic Melt

Before we come to the filling study in Chapter 6 "Filling study, holding pressure and mold clamping force", it is necessary to check the following points:

- The temperature of the melt
- The appearance of the melt

5.7.1 Checking the Melt Temperature

The temperature of a plastic melt ejected from the barrel is measured with a thermometer (Figure 5.23). This allows you to judge whether the processing temperature recommended by the raw material manufacturer obtains. The melt temperature also provides interesting information about the additional heat generated by the friction introduced during the plasticizing process.

5.7.2 Check the Melt Appearance

The molded plastic melt (Figure 5.24) must be visually assessed. The following questions should be examined:

- Is the plastic homogeneous and completely melted?
- Is there any thermal damage in the melt, e.g. brownish discoloration?
- Are there any bubbles in the melt?
- Are there foreign particles, such as metal chips, recognizable in the melt?
- Are there any color differences in the melt?

Figure 5.23 Checking the melt temperature

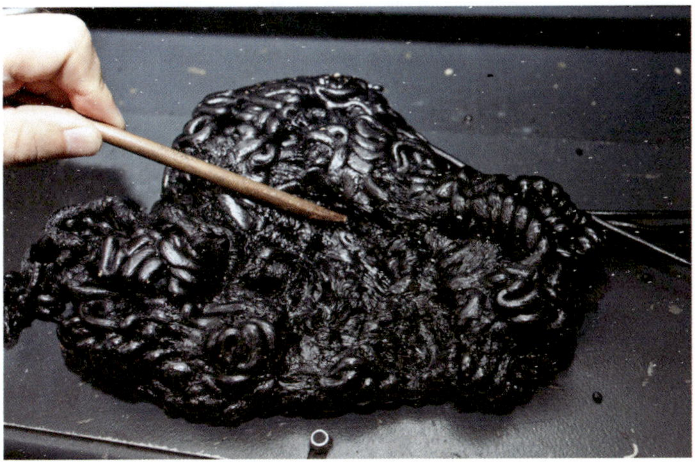

Figure 5.24 Molded plastic melt

5.7.3 Checking the Mold Wall Temperature

The last point before the filling study is to measure the temperature of the mold and the temperature difference between the flow and return with a thermometer (Figure 5.25). The mold must have reached the required temperature and the measured difference between the flow and return should not be more than 2–3 °C.

Figure 5.25 Measuring the nozzle-side and ejector-side mold wall temperature

Milestone: "Basic setting of the plasticizing unit"

- Block 4 of the mold sampling checklist has been processed in full
- All relevant basic settings of the plasticizing unit have been made
- Nozzle impression has been created
- The melt has been checked
- The basic setting of the plasticizing unit has been completed

References

[1] Schröder, T.: Rheologie der Kunststoffe- Theorie und Praxis, Carl Hanser Verlag, Munich, 2018

[2] Degener, C., Camatta

[3] Johannaber, F., Michaeli, W.: Handbuch Spritzgießen, 2nd edition, Carl Hanser Verlag, Munich, 2004

6 Filling Study, Holding Pressure and Mold Clamping Force

This chapter will deal with the filling study first, then with the holding pressure and finally with the mold clamping force. It describes how you carry out a filling study and the important findings that emerge from it. It will also show you how to set the optimum holding pressure, the correct switchover point from injection to holding pressure and the holding pressure time required for your molded part. At the end, I would like to show you how to determine the required clamping force for your injection mold and what information you can obtain about the mold from this.

In addition to this chapter, Block 6 ("Filling study, holding pressure and mold clamping force") from the mold checklist (Figure 6.1) is of relevance for the "mold sampling specialist".

Filling study, holding pressure and mold clamping force				
Assessment of the plastic melt:		**Yes**	**No**	**Block 6**
29	Melt temperature and mold wall temperature checked?			
30	Assessment of the melt carried out?			
Filling study, holding pressure and mold clamping force:		**Yes**	**No**	
31	Filling study (up to 100% volumetrically filled parts) produced and holding pressure level optimized?			
32	Filling study evaluated and documented?			
33	Required holding pressure time (sealing time) determined and documented via the molded part weight?			
34	Required clamping force (kN or t) optimized with dial gauge?			

Figure 6.1 Mold checklist: Block 6

6.1 Filling Study

A filling study (Figure 6.2) shows how the flow front of the molded part advances over time during the injection phase. This provides the "mold sampling specialist" with important information about the molded part, the mold, the basic setting of the process parameters and the required switchover point from injection to holding pressure.

Figure 6.2 Example of a filling study

6.1.1 Findings from the Filling Study

6.1.1.1 Knowledge about the Molded Part and the Mold

A filling study enables you to examine the formation (flow fronts and flow paths) of the molded part in self-defined partial fillings and thus identify critical areas, such as

- Confluence points (flow and weld lines)
- Air inclusions
- Changes in wall thickness in the molded part, and
- Uneven filling of multiple cavities

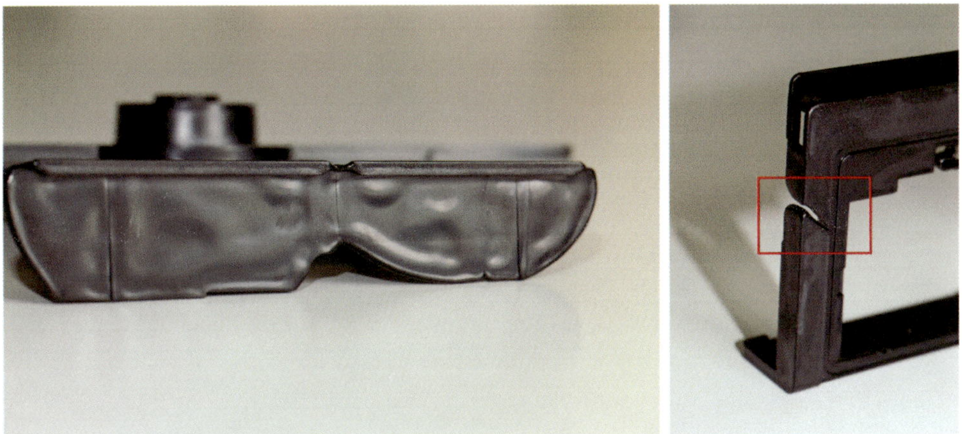

Figure 6.3 Example of confluence points (weld line)

The filling study provides the following information about the mold:

- Functionality of the mold vents installed
- Critical areas that require additional venting, e.g. lamella pack, porous sintered insert, vent plugs, etc.
- Balanced or even filling of the molded part

Figure 6.4 Example of balanced filling

6.1.1.2 Knowledge about the Basic Setting of the Process Parameters

In Chapter 5 "Basic settings of the plasticizing unit", you determined an initial basic setting of the process parameters for your molded part. Using the filling study, it is now possible to perform an initial check or optimization of the following basic settings for the process parameters.

Required shot volume

(see Section 5.2.1.1 "Determining the required plasticizing volume or plasticizing stroke")

You can determine the actual shot volume required for your molded part by employing a range of defined fillings, from partial through to fully molded.

Injection speed profile

(see Section 5.3.2 "Injection profile")

During the filling study, you can determine in which areas of the molded part there are critical areas, such as changes of wall thickness. In order to achieve a constant injection speed here, all filling patterns with changes of wall thickness must be marked with the instantaneous position of the plasticizing screw (mm or cm^3). This allows you to create a customized injection speed profile based on the position information.

Demolding temperature of the molded part

(see Section 5.4.2 "Demolding temperature of the plastic")

During the filling study, the "mold sampling specialist" must check whether the required demolding temperature has been reached. This can be checked with a thermometer or with a thermal imaging camera on the surface of the molded part. If molded parts have a higher demolding temperature, the residual cooling time must be adjusted.

6.1.2 Procedure for the Filling Study and Determination of the Switchover Point or Volume

In the following section, I would like to show you how you can create a filling study and at the same time determine the correct switchover point or the correct switchover volume for your molded part. The procedure is shown step by step using the "Move switchover point" method.

Filling study

- **Step 1:** Holding pressure and holding pressure time must be switched off or set to "zero" and the injection molding machine must be switched from set-up mode to semi-automatic mode

- **Step 2:** The theoretical plasticizing volume (shot volume + residual melt cushion; see Section 5.2.1.1 "Determining the required plasticizing volume or plasticizing stroke") is set on the machine and plasticized
- **Step 3:** For the start of the first partial filling, the switchover point or the switchover volume must be set so large that over-molding of the mold is avoided. It is also important to ensure that the first partial filling is large enough to safely demold the molded part. It is very difficult to define a general start for the switchover point, as this depends on the molded part. As an initial value, you should set the first switchover point at 10–20% before the set plasticizing volume in the direction of the machine nozzle
- **Step 4:** The switchover point or the switchover volume is now shifted in 5–10% steps in the direction of the machine nozzle. However, the step size must be adapted to the complexity of your molded part. This allows you to obtain more precise information about critical areas of the molded part by using smaller increments in the filling study

NOTE:

In order to be able to draw precise conclusions from the filling study, all partial fillings must be labeled with the respective set switchover point or volume (in mm or cm^3)!

Switchover point or volume

- **Step 5:** The filling study is successfully completed when the molded part has been filled 100% volumetrically. This instantaneous screw position (in mm or cm^3) at 100% molded part filling is also our required switchover point from injection pressure to holding pressure

REMEMBER:

Switchover at 100% volumetric molded part filling is very important, as residual filling of the molded part via holding pressure constitutes an uncontrolled process. Filling the molded part via holding pressure can have the following negative effects: higher mold loads (wear), over-loaded molded part and higher internal stresses.

- **Step 6:** The last step is to check the residual melt cushion in the screw antechamber (see Section 5.2.1.1 "Determining the required plasticizing volume or plasticizing stroke"). There should be enough residual melt cushion at the end of the filling study for the upcoming holding pressure phase. If, for example, the residual melt cushion is inadequate (< 10 mm), you can optimize it using the following practical tip

PRACTICAL TIP:

If the existing residual melt cushion is inadequate after 100% volumetric filling, you can adjust it on the machine with these simple steps:

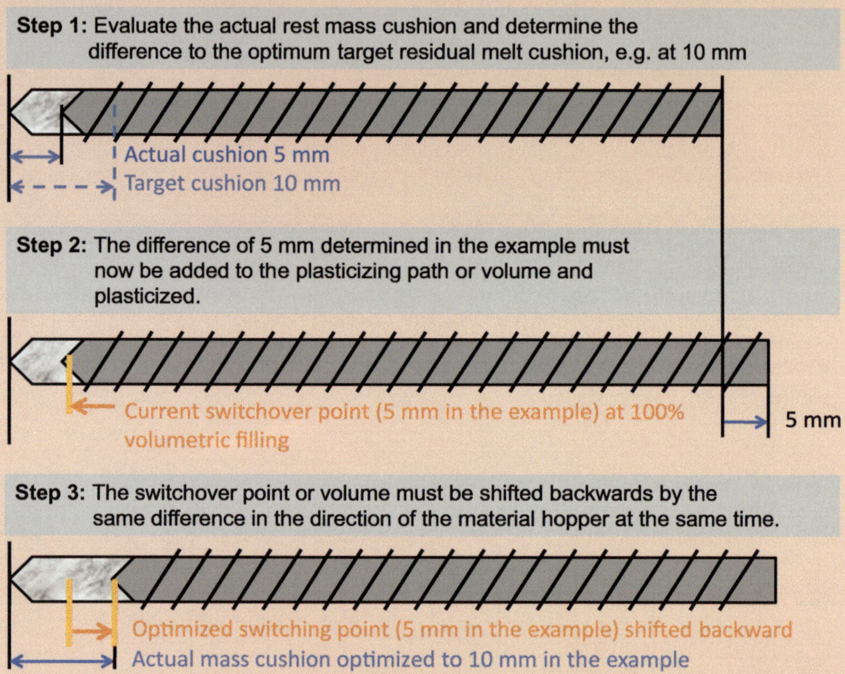

This method ensures that the optimum switchover point is still reached at 100% volumetric filling and that the injection mold cannot be over-loaded.

You can apply these steps in reverse for a melt cushion that is too large. This means that you shorten the plasticizing stroke and at the same time shift the switchover point in the direction of the machine nozzle by the same amount.

6.1.3 Types of Switchover from Injection Pressure to Holding Pressure

Once you have determined the switchover point or the switchover volume of your molded part using the filling study, you must now determine the type of switchover from injection pressure to holding pressure. To do this, I would first like to briefly introduce you to the possible types of switchover in injection molding.

Stroke- or volume-dependent switchover (mm or cm^3)

Most injection molding companies work with stroke- or volume-dependent switchover, as this is easy to implement and usually sufficiently accurate. The switchover signal takes place via a defined screw position and is triggered by a limit switch or a position sensor. With this type of switchover, a constantly reproducible end of the plasticizing stroke or plasticizing volume as well as a consistent closing behavior of the non-return valve must be guaranteed during the process.

With hydraulic machines, the inertia of the hydraulic drive system can cause the defined switchover position of the plasticizing screw to be overrun and the switchover signal to arrive too late. This usually leads to pressure peaks in the pressure curve in the mold at very high injection speeds. For this reason, care should generally be taken in the case of stroke-dependent switchover to ensure that the injection speed is significantly reduced directly before the switchover position by using a stepped injection profile. This allows you to achieve improved switchover accuracy. Given an optimized machine setting, the fluctuation of the switchover point should not exceed ± 0.1 mm.

Time-dependent switchover (s)

Time-dependent switchover has not proved very successful and is therefore very rarely used. In this case, an injection time must first be determined via the stroke-dependent switchover until volumetric filling of the molded part has been achieved. The switchover signal is sent when a defined injection time has elapsed. With this type of switchover, a consistently reproducible screw end position after plasticizing is not relevant. However, the closing behavior of the non-return valve should be consistent. The major disadvantage of this method is that it is highly dependent on the injection speed, the plasticizing accuracy, the mode of operation of the non-return valve, material fluctuations and temperature fluctuations. Nowadays, speed-controlled machines achieve sufficient switchover accuracy by this method.

Hydraulic pressure-dependent switchover (bar)

Switchover via the hydraulic pressure is also very rarely used. With this method, the switchover signal is given by a pressure sensor attached to the hydraulic injection piston of the injection molding machine. This pressure sensor triggers the switchover to holding pressure at a defined pressure level.

With this switchover method, the inertia between the oil and the melt can be neglected, in contrast to stroke-dependent switching, as switching usually has to take place before the pressure signal is reached when your molded part is filled volumetrically. One disadvantage is that different melt temperatures, for example, cannot be recognized by this switchover method–as is the case for cavity pressure-dependent switchover presented below.

Cavity pressure-dependent switchover (bar)

Cavity pressure-dependent switchover is the most precise method of switching from injection pressure to holding pressure. With this method, a pressure signal for switching over is transmitted to the injection molding machine via a pressure sensor integrated in the mold cavity close to the sprue. For this purpose, the cavity pressure during the volumetric filling of your molded part must be determined experimentally via the filling study or with the aid of an injection molding simulation. A prerequisite for switching via the cavity pressure is for the injection molding machine to have a dynamic hydraulic system and a fast electrical control system. This is very important, as the cavity pressure rises very quickly during volumetric filling of the molded part and precise switching is otherwise not possible.

The advantage of cavity pressure-dependent switchover is that all uncontrollable fluctuations, such as

- Fluctuations in the plasticizing stroke
- Differences in the density of the plastic
- Changes in flow properties, such as viscosity fluctuations
- Changes in mold and material temperatures, and
- Changes in injection speeds

have virtually no influence on the switchover accuracy, as this switchover only ever takes place at a defined internal pressure, i. e. during volumetric molded part filling.

As good and precise as this switchover method is, there are also disadvantages, such as:

- Higher costs, due to internal pressure sensors required in the mold
- Higher machine costs for interfaces and programs
- Marking of the pressure sensor on the molded part
- Susceptibility of the measuring system (pressure sensor) to faults, as it is housed in the mold and is not part of the injection molding machine

With the help of the cavity pressure curve, it is possible to draw conclusions about the optimum switchover point as well as other important information about the process. For this reason, all important information on cavity pressure is explained in detail later in connection with the mold sampling analysis (Chapter 7 "Mold sampling analysis of the basic setting").

6.1.4 Effects of the Switchover on the Molded Part and the Injection Process

Correct and precise switchover from injection pressure to holding pressure has a major influence on the quality of your molded part. I would therefore like to give you an overview of how switching too early or too late can affect the molded part and the injection molding process.

Table 6.1 Effects of Switchover on the Molded Part and the Injection Process

Switching to holding pressure too early means:
Insufficient filling of the molded part
Sink marks
Weld line markings
Higher shrinkage potential
Lighter molded parts
Switchover marks on the molded part, due to premature stopping of the flow front
Switching to holding pressure too late means:
High pressure peaks in the pressure curve
Over-molding or flash formation on the molded part
Molded part dimensions too large
Heavier molded parts
Increased orientation in the molded part in the area of the gate
Increased clamping force required by the injection mold
Increased stresses in the molded part

6.2 Holding Pressure

Once the filling study has been completed and the switchover point determined, the holding pressure level (bar), holding pressure time (s) and, in special cases, the holding pressure profile (mm/s) are set. In the following sections, all the important background information and procedures for setting are shown using practical examples. First, I would like to explain the meaning of holding pressure.

BACKGROUND KNOWLEDGE:

What is the holding pressure needed for?

At a large number of courses or in technical books you tend to hear or read: "Holding pressure is there to compensate for volume contraction or volume shrinkage after the plastic has been injected". Such statements are completely correct, but I would also like to explain to you what the volume shrinkage is in general which you are trying to compensate for every day on the injection molding machine with the help of holding pressure.

Let's start with the initial situation in the injection molding process. The polymer is transformed into a plastic state by heating. This increases the volume of the polymer. This increase in volume can be visualized with the help of a cup model (Figure 6.5). Imagine that a cavity with a mold volume (V_{WZ}) of 0.955 cm^3 has been prepared. The 1 g of polystyrene (PS) in the left-hand cup would fit exactly into this cavity at room temperature and atmospheric pressure. If the PS is now heated so that it becomes plastic and therefore flowable and moldable, the volume of the PS increases. The consequence of filling without pressure would be that the 1 g of PS would not fit completely into the cavity, leaving a residue in the cup. At the end of the cooling time, exactly this amount remaining in the cup would be missing in the cavity. This shortfall in relation to the mold volume is therefore known as volume shrinkage. The volume shrinkage would be in the order of approx. 10% for unpressurized processing of amorphous polymers and approx. 25% for semi-crystalline polymers. Such molded parts would be severely warped and have a wavy and uneven surface as well as sink marks. These molded parts would be pure rejects and unusable. In order to counteract this volume shrinkage, part of the volume increase must therefore be compensated for by compression with the holding pressure after injection [1].

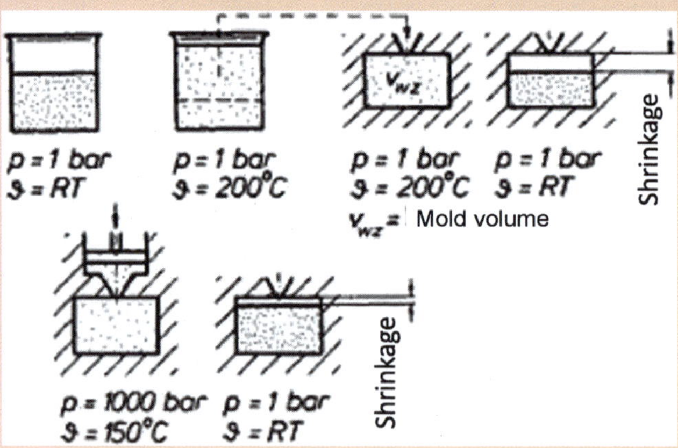

Figure 6.5 Cup model for molded part formation (image source: Stitz S., Keller W.: Spritzgießtechnik, 2nd edition, Carl Hanser Verlag, Munich Vienna, 2004, page 90, image number 1.67)

6.2.1 Holding Pressure Level

In the injection molding industry, it is often said that the holding pressure should be 40–60% of the required specific injection pressure. This recommended value range is based on empirical values obtained from test series conducted on the injection molding machine. In principle, these values can be taken as an initial guide to the required holding pressure level, but you should always determine the required holding pressure level of your molded part yourself by carrying out experiments.

Before the optimum holding pressure level is addressed, Table 6.2 summarizes all the effects on the molded part and the injection molding process that arise from an increase in holding pressure.

Table 6.2 Effects of Increasing the Holding Pressure

Effects on the molded part:
Lower volume contraction (volume shrinkage) in the molded part
Possible demolding problems
Larger molded part dimensions
Higher molded part weight
Stronger weld lines
Risk of flash formation
Greater orientation
Avoidance of sink marks and blowholes
Higher internal stresses
Reduced warp
Effects on the injection molding process:
Greater clamping force required for the mold
Risk of molded part demolding under residual pressure

6.2.1.1 Procedure for Determining the Required Holding Pressure

The level of holding pressure required for your molded part is determined experimentally at the injection molding machine as follows:

Starting point: Step 6, Section 6.1.2 "Procedure for the filling study and determination of the switchover point or volume": The inspection or optimization of the melt cushion has been completed and there is a sufficient cushion in the screw chamber.

- **Step 1:** Switch on the holding pressure on the injection molding machine
- **Step 2:** Start with a very low holding pressure level, e. g. at 10–20% of the specific injection pressure, and visually inspect the molded part for wavy surfaces and sink marks

> **NOTE:**
> A longer holding pressure time must be preset so that the holding pressure can take effect in the molded part. This is then optimized under holding pressure time. Take particular care to ensure that the holding pressure level is set to a minimum during the first attempts. → Otherwise there is a risk of over-loading the mold, damage to the mold or over-molding!

- **Step 3:** Slowly increase the holding pressure level from molded part to molded part in increments
- **Step 4:** The optimum holding pressure level is reached when the molded part is not over-molded and has no wavy surfaces or sink marks
- **Step 5:** Finally, the residual melt cushion must be checked and optimized if necessary (see practical tip in Section 6.1.2 "Procedure for the filling study and determining the switchover point or volume")

6.2.2 Holding Pressure Time

The holding pressure time is the length of time that the holding pressure can act in your molded part to compensate for the volume shrinkage. The holding pressure time depends on the duration of the sealing point, which indicates when the gate or the connection to your molded part is solidified or sealed. When the sealing point is reached, no melt movement and therefore no pressure transfer to your molded part is possible. It is therefore crucial to know or determine the time of the sealing point in order to avoid unnecessarily extending your cycle time.

The time it takes to reach the sealing point depends on several factors, such as

- Melt and mold temperature
- Type of gating system (cold gating, hot gating via hot runner), and
- Cross-Section of the gating

If a high mold temperature or a high melt temperature is present, the time required to seal the gate is extended and holding pressure can be maintained in the melt for longer. With a cold runner gate, the sealing point is reached when the gate has solidified across its entire cross-section. With a hot runner gate, on the other hand, the sealing point is reached when the melt has solidified at the first cross-section in the gate area of the molded part.

The sealing point can be determined by the following methods:

- Computer simulation programs
- Analytically via a cooling equation
- Cavity pressure
- Experimentally via the molded part weight

Experience has shown that determining the sealing time using simulation programs or cooling equations is a sensible approach for making initial estimates. These two methods require experience in FEM calculations on the one hand and a greater knowledge of mathematics on the other.

Determining the required holding pressure time				
Article description:				
Mold number:				
Date:				
Created by:				
Shot-nr.	**Holding pressure time(s)**	**Rest cooling time (s)**	**Molded part weight (g)**	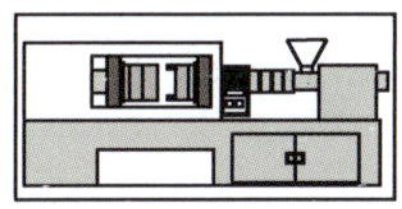
1				
2				
3				
4				
5				
6				
7				
8				
9				
10				
11				
12				
13				
14				
15				
16				
17				
18				
19				
20				
21				
22				
23				

Figure 6.6 The "Determining the required holding pressure time" template

One problem with the cavity pressure method is that many molds are not equipped with pressure sensors. For this reason, I would like to explain the most frequently used method in injection molding companies, the method of using the weight of the molded part. The reason for this is that it is easy to perform on the machine.

Unfortunately, molding shops often use different and sometimes incorrect procedures to determine the sealing point by the molded part weight method. In order to ensure a correct and standardized approach, a template for determining the required holding pressure time is shown in Figure 6.6.

The next section uses this template to show you how to determine the required holding pressure time step by step.

6.2.2.1 Procedure for Determining the Required Holding Pressure Time

Starting point: All basic settings for the process parameters, except the holding pressure time, have been completeded. The experimentally determined holding pressure level (Section 6.2.1.1 "Procedure for determining the required holding pressure") has been set.

- **Step 1:** The "mold sampling specialist" takes the template for determining the required holding pressure time (Figure 6.6) from the mold sampling documents
- **Step 2:** Reduce the preset holding pressure time again and start to produce a molded part with a very low holding pressure time setting, e.g. 0.5 s. The commenced holding pressure time must be documented in the first line of the template. Keep the level of the holding pressure you have already determined constant
- **Step 3:** Now subtract the holding pressure time with which you start the sealing point determination from the set residual cooling time (see Section 5.4.3 "Basic setting of the residual cooling time") and enter the result under Residual cooling time in the template

> **REMEMBER:**
>
> As the cooling time consists of the holding pressure time and the residual cooling time, the residual cooling time must be reduced by the same amount of time as the holding pressure time is increased. This is the only way to ensure a consistent cycle time.

- **Step 4:** The manufactured molded part is weighed without sprue and the molded part weight is documented in the first line of the template
- **Step 5:** Continue to increase the holding pressure time per shot in small steps (e.g. 1 s, 1.5 s, 2 ...) and weigh the molded part
- **Step 6:** The sealing point of your molded part is reached (see Figure 6.7) when there is no more weight gain and the weight remains constant

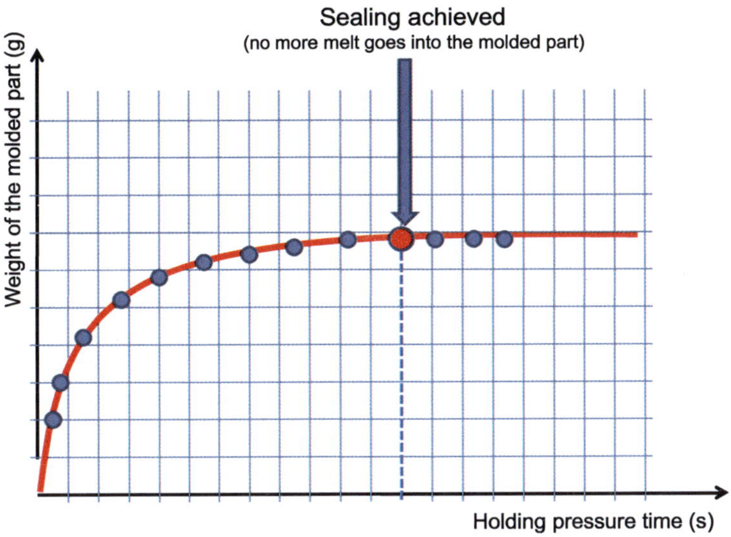

Figure 6.7 Determination of the sealing point via the molded part weight

6.2.3 Holding Pressure Profile

In practice, a one-step holding pressure profile (Figure 6.8) with a constant holding pressure level is usually set. The reason for this is that it is very difficult for the "mold sampling specialist" to evaluate the results of the set profile. As different residual stresses and orientations in the molded part are influenced by a profile, this cannot be measured on the machine by simple methods and is therefore not reproducible.

However, aiming for a holding pressure profile is fundamentally correct and is sometimes necessitated by the molded part. In order to be able to set the correct holding pressure profile, the differences between amorphous and semi-crystalline thermoplastics must first be considered in more detail.

With amorphous thermoplastics, the cavity pressure during the holding pressure time drops in line with the falling molded part temperature. This is due to the ever-increasing viscosity of the melt and the associated poorer pressure transfer from the screw antechamber into the mold until the ambient pressure of 1 bar is reached.

As a result, high internal stresses and orientations usually occur near the gate area to the molded part and lead to distortion and stress cracking of the molded part. To avoid this, it is advisable to set a decreasing holding pressure profile (Figure 6.9) for amorphous thermoplastics.

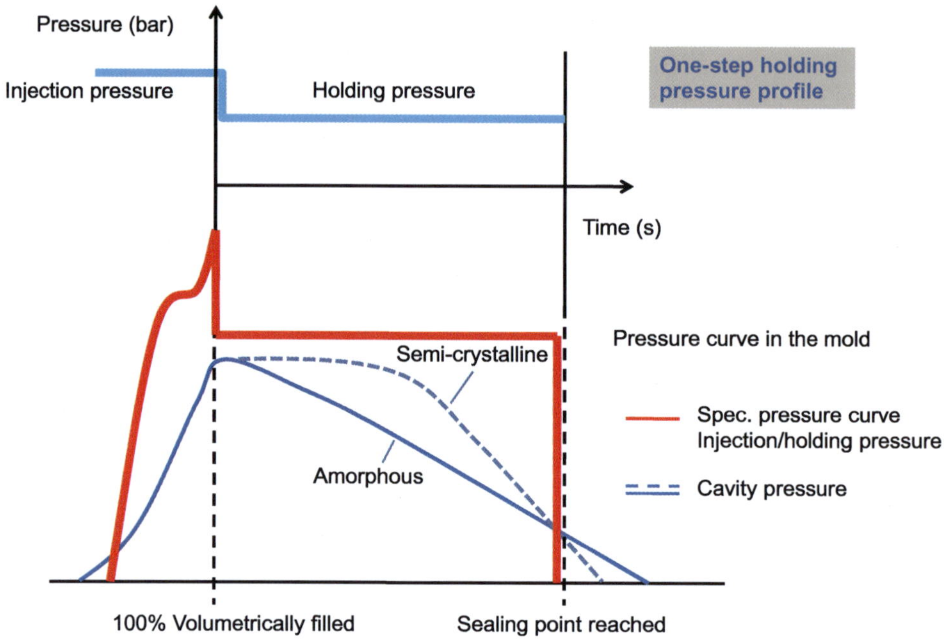

Figure 6.8 One-step holding pressure profile

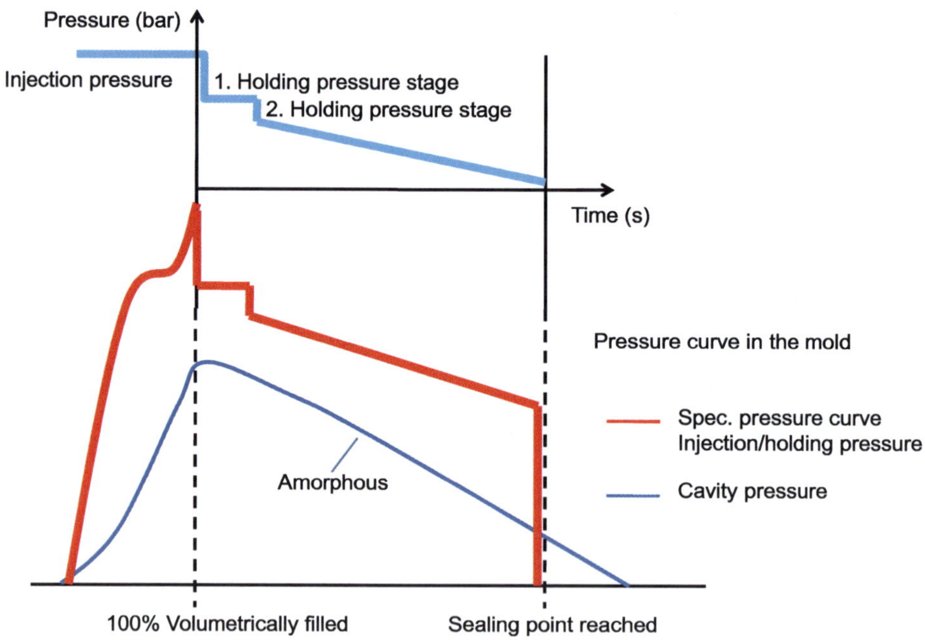

Figure 6.9 Falling holding pressure profile for amorphous thermoplastics

With semi-crystalline thermoplastics, there is an almost constant cavity pressure during the holding pressure time until the crystallization temperature is reached, due to better pressure transfer. Subsequently, due to the strong volume shrinkage in the solid area during crystallization, there is a steep drop in pressure to the ambient pressure of 1 bar.

In the case of semi-crystalline thermoplastics, up to 10–15% of the melt must be added during the holding pressure phase, due to the high shrinkage. The viscosity initially remains almost the same inside the mold and increases abruptly when the temperature falls below the crystallite melting temperature. If the pressure load is changed by the holding pressure during this cooling time, the crystallization also changes and becomes more uneven. These fluctuations can in turn have a negative effect on the molded part properties. A constant holding pressure profile (Figure 6.10) until the sealing point is reached is therefore recommended.

NOTE:

After a holding pressure profile has been set, the residual melt cushion must always be checked or optimized if necessary (see practical tip in Section 6.1.2 "Procedure for the filling study and determining the switchover point or volume").

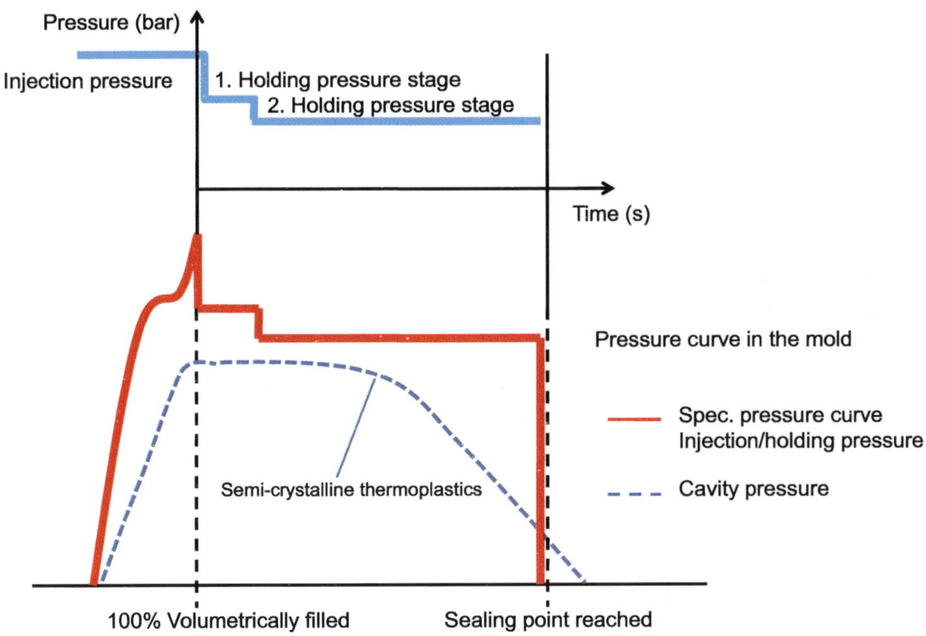

Figure 6.10 Constant holding pressure profile stepped for semi-crystalline thermoplastics

6.3 Mold Clamping Force

Nowadays, it is increasingly common practice to operate with the maximum possible clamping force of the respective injection molding machine or a to apply a force that has only been theoretically determined using approximation formulas. As a result, excessive clamping forces can lead to high surface pressure, which promotes wear on your mold as well as venting problems. Excessive clamping force also has a negative effect on the energy efficiency of your process, as unnecessarily high clamping forces require more energy.

In order to arrive at the optimum clamping force for your sampling mold, it is necessary to perform an experimental optimization of the theoretical clamping force already set on the machine from Section 4.3 "Basic setting of the mold clamping force".

Before we get to the experimental optimization of the clamping force, I would like to provide you with important information on the subject of "mold clamping force" in the following "Background knowledge" information box.

BACKGROUND KNOWLEDGE:

The clamping force (kN) is the force that must counteract the buoyancy force which is generated in the mold during the injection and holding pressure phases and results from the maximum cavity pressure (pressure peak) and the projected area of the molded part.

The buoyancy force ensures that the volume of the cavity in the mold increases in the closing direction. This phenomenon is also referred to as "mold breathing" and is dependent on factors such as:

- The level of cavity pressure
- The rigidity of your injection molding machine
- Rigidity of the mold
- Type of clamping system employed on the machine

An internal mold pressure of 450 bar causes a compressive force of 900 kg on an area of 2 cm^2 or the size of a fingernail. Using this example, you can see that huge forces in the cavity during injection and holding pressure are responsible for mold breathing and thus deformation of the mold. Figure 6.11 shows a schematic representation of the deformation of a mold under cavity pressure.

Deformation can assume the following quality characteristics for the molded part and your injection molding process:

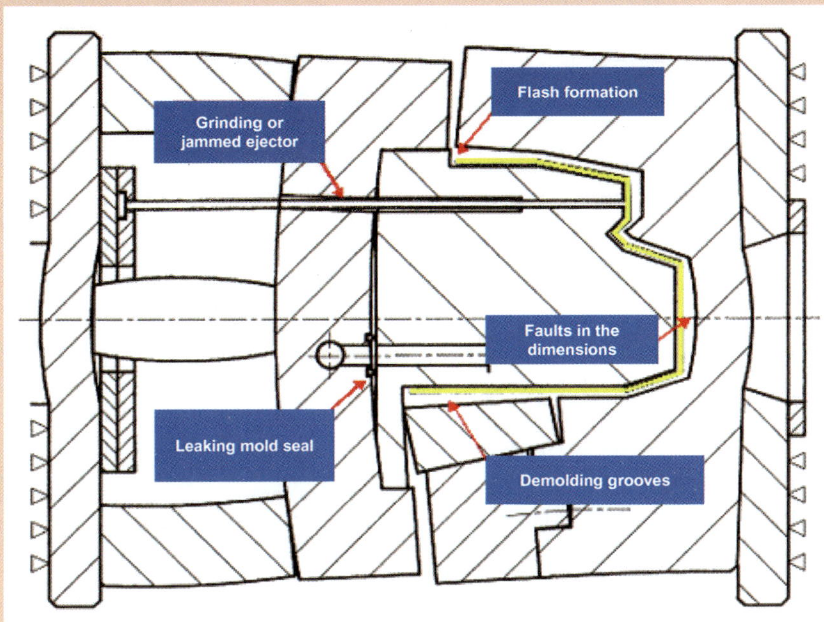

Figure 6.11 Deformation of a mold under cavity pressure (schematic) [1]

- **Dimensional fault:** Strong mold breathing can lead to an increase in the dimensions of the molded part. Furthermore, spring-back of the mold due to breathing creates a compression effect, which can influence the dimensions of the molded part, such as an uncontrolled holding pressure
- **Flash formation, skin, unclean separation:** As can be seen in Figure 6.11, the mold platens deform in an axial direction. As a result, the sealing effect in the parting line is canceled and the melt can penetrate
- **Demolding difficulties, surface defects on the molded part:** If the deformations in the radial direction exceed the wall thickness shrinkage of the molded part, the molded part may become jammed by the spring-back mold as if in a vise, which leads to possible demolding under residual pressure. The consequences are tilting of the molded part during demolding, which leads to scratch marks on the molded part in the demolding direction
- **Wear (seizing) or tilting of the ejector system in the mold:** If the mold plates are deformed too much, the ejector pins can no longer be guided and centered properly. This leads to sluggishness of the ejectors and may even cause them to seize up
- **Leaks:** As the cooling circuits sometimes run over several platens of the injection mold, these are sealed via the platen pressure and additionally with sealing elements, such as sealing rings. If the platens are deformed too much, the sealing function becomes ineffective and the cooling medium escapes

- **Offsetting of the mold inserts:** Unequal pressure loads in the mold during the injection and holding pressure phase can lead to core offset. This is responsible for problems during filling as well as for wall thickness errors and demolding problems with the molded part
- **Inefficiency:** Unacceptable mold deformation accelerates mold wear considerably, which leads to increased maintenance and repair costs. As this wear in the mold is gradual, it causes quality problems in your molded part, which in turn leads to further costs, such as rejects and complaints [1]

6.3.1 Experimental Optimization of the Mold Clamping Force

This section deals with adapting the basic setting of the clamping force to your injection process by means of an experimental procedure at the machine. The "Experimental optimization of the mold clamping force" template (Figure 6.12) has been developed to help you implement the injection tests. In addition to the template, you will also need a dial gauge (accuracy: 1/100 mm) and a balance (accuracy: 0.01 g).

Experimental optimization of the mold clamping force		
Article description:		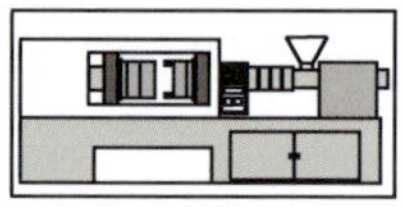
Mold number:		
Date:		
Created by:		
Clamping force(❐ kN ❐ t)	Mold breathing measured (mm)	Weight of the molded part (g)

Figure 6.12 The "Experimental optimization of the mold clamping force" template

REMEMBER:

The permissible mold breathing depends on three factors: the general quality requirements for the molded part, the plastic to be processed and the size of the injection molding machine. Practical experience has shown that the maximum mold breathing for small injection molding machines (up to 1500 kN) is approx. 0.01 mm and for medium-sized machines (up to 3500 kN) up to approx. 0.02 mm. For larger injection molding machines (> 3500 kN), mold breathing can be 0.06–0.1 mm.

6.3.1.1 Method for Experimental Optimization of the Mold Clamping Force

The "mold sampling specialist" takes the "Experimental optimization of the mold clamping force" template from the sampling documents.

Starting point: The machine setting parameters such as holding pressure, holding pressure time, holding pressure switchover, injection pressure and injection profile have been set.

- **Step 1:** The dial gauge must be attached to the mold (see Figure 6.13) and zeroed when the clamping force is applied

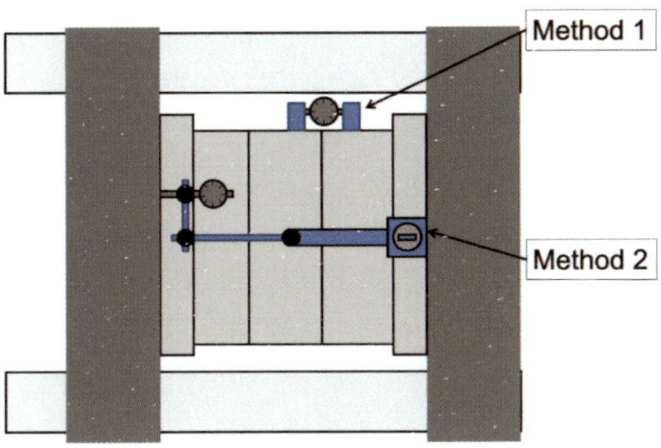

Figure 6.13 Possible arrangements of the dial gauge on the mold

- There are two methods for setting the dial gauge:
  - **Method 1:** Between the nozzle-side and ejector-side mold plates directly on the mold → this means you can only measure the mold breathing as a function of the buoyancy force
  - **Method 2:** Between the clamping platens on the side of the mold, in the middle between the machine beams → this allows you to measure mold compression as a function of the clamping force and breathing of the mold as a function of the buoyancy force

- **Step 2:** At the beginning of the experimental test, it is not the maximum clamping force of your injection molding machine that is taken as the initial value, but rather the clamping force you have already determined (Section 4.3 "Basic setting of the mold clamping force")
- **Step 3:** Now produce a molded part and read the mold breathing (plate movement 0.01–0.02mm) on the dial gauge during the injection phase and document this in the "Experimental optimization of the mold clamping force" template
- **Step 4:** The manufactured molded part is labeled with the clamping force currently set and then weighed. The molded part weight must also be documented in the template
- **Step 5:** Now reduce the clamping force slightly and produce another molded part. To do this, repeat steps 3 and 4
- **Step 6:** Perform step 5 until the injection mold has reached the maximum permissible mold breathing
- **Step 7:** Visually check your molded part for flash. If it is present, you have already fallen below the minimum clamping force required. In this case, the clamping force must be adjusted upward until there is no more flash on your molded part

NOTE:

In order to be able to correctly assess if the flash on your molded part is within the permissible range, you should read the permissible value of the maximum parting line from the molded part drawing, if available.

- **Step 8:** As mold breathing can only be measured completely and precisely if breathing is measured simultaneously on all four sides of the mold, you must add a safety factor of 10–15% to the minimum required clamping force to ensure safe and trouble-free injection molding

PRACTICAL EXAMPLE:

Using the practical example of the PC/ABS housing cover, I will show you how to determine the optimum clamping force using the "Experimental optimization of the mold clamping force" template.

Experimental optimization of the mold clamping force		
Article description:	Housing cover	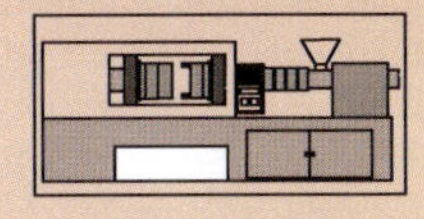
Mold number:	11082006	
Date:	XXXXXXXX	
Created by:	XXXXXXXX	

Clamping force(☒ kN ☐ t)	Mold breathing measured (mm)	Weight of the molded part (g)
3500	0	261.54
3300	0	261.72
3200	0.01	261.72
2900	0.01	262.01
2800	0.02	262.53
2750	0.03	262.94
2550	0.04	263.07

Figure 6.14 Practical example of "Experimental optimization of the mold clamping force"

The case study shows very clearly that mold breathing of approx. 0.02 mm was achieved with a clamping force of 2800 kN. The next step here is to add a safety factor of 10–15%. In our example, you therefore need a clamping force of approx. 3080 kN.

6.3.1.2 Knowledge Gained from the Experimental Optimization of Mold Clamping Force

This experimental procedure not only provides the clamping force required for your molded part, but also other important knowledge about the rigidity of your injection mold. The knowledge gained can be as follows:

- **"Soft" mold:** A "soft mold" is present if the weight of the molded part increases even though no mold breathing can be detected on the dial gauge. Consequently, this shows that only deformation of the cavity takes place. The clamping force is sufficient in this case and cannot be determined more precisely

- **"Hard" mold:** A "hard mold" is present when an increase in the weight of the molded part occurs with measurable mold breathing detected on the dial gauge. In this case, the boundary conditions are optimal and the required clamping force can be determined precisely
- **Constant "mold breathing":** Breathing that remains constant after the holding pressure phase indicates possible over-molding or flash in the parting line. This case is problematic and must be optimized, as a very high surface pressure occurs on the spotting surfaces in the parting line and promotes wear of the mold

Milestone: "Filling study, holding pressure and mold clamping force"

- Block 5 of the mold sampling checklist has been processed in full
- The filling study and the determination of the switchover point to holding pressure have been carried out
- Holding pressure has been selected as the type of switchover
- The required clamping force has been optimized at the machine by an experimental procedure

References

[1] Stitz, S., Keller, W.: Spritzgießtechnik, 2nd edition, Carl Hanser Verlag, Munich Vienna, 2004

7 Mold Sampling Analysis of the Basic Setting

The “mold sampling specialist” systematically works through Block 7 (“Mold sampling analysis of the basic setting”) of the mold sampling checklist (Figure 7.1).

Manufacturing first tooling samples and conducting a sampling analysis of the basic setting				
Producing the first tooling samples:		**Yes**	**No**	
35	Machine capability in accordance with VDMA 24.470 Parts 1 and 2?			
36	Cavity pressure sensors connected to the machine?			
37	Injection molding process run in over several cycles (20–30)?			
Sampling analysis of the basic setting:		**Yes**	**No**	
38	Analysis of the required flow rate of the temperature control medium carried out?			
39	All faults occurring on the molded part localized and defined?			**Block 7**
40	Cause of error on the molded part localized and defined?			
41	Specific analysis of injection molding process and mold carried out?			
42	**General/visual inspection of the process:**			
43	Target/actual comparisons of all machine setting parameters carried out?			
44	Is the existing residual melt cushion OK and documented in the machine setting data sheet?			
45	Is the dwell time of the melt in the barrel calculated, checked and documented in the machine setting data sheet?			

Figure 7.1 Mold sampling checklist: Block 7

Manufacturing first tooling samples and conducting a sampling analysis of the basic setting				
Producing the first tooling samples:		**Yes**	**No**	
46	**Thermal checks (thermography):**			
47.1	Thermal pictures or videos (thermography) taken?			
47.2	Demolding temperature (target/actual comparison) OK?			
47.3	Temperature distribution in the mold and molded part OK?			
47.4	Temperature difference between flow and return temperature of the temperature control unit measured and OK?			
47.5	Is the cavity temperature homogeneous?			
47.6	No hot spots in the mold or on the molded part?			
48	**Cavity pressure curve:**			
48.1	Cavity pressure curves recorded and checked?			
48.2	Injection phase optimal?			
48.3	Compression phase optimal?			
48.4	Correct switching point for holding pressure?			
48.5	Holding pressure level and holding pressure time optimal?			
48.6	Not demolded under residual pressure?			
49	**Evaluation of the molded part and the mold:**			
49.1	Molded part free of ejector marks?			**Block 7**
49.2	Correct mold cavity codes?			
49.3	Date clock and index set correctly?			
49.4	Molded part free of flash?			
49.5	Molded part free of ejector marks?			
49.6	Ejector not protruding?			
49.7	Ejector not retracted?			
49.8	Molded part fully injected?			
49.9	Are the installed mold vents functional?			
49.10	Are the dimensions of the molded part OK?			
49.11	Molded part without bubbles/air bubbles/air pockets?			
49.12	Molded part without warpage?			
49.13	Molded part without sink marks?			
49.14	Molded part without streaks?			
49.15	Molded part without scratches or marks?			
49.16	Molded part without stress cracks (white crack)?			
49.17	Molded part without shiny spots?			
49.18	Molded part without matt spots?			

Figure 7.1 Mold sampling checklist: Block 7 (*continued*)

Manufacturing first tooling samples and conducting a sampling analysis of the basic setting				
Producing the first tooling samples:		**Yes**	**No**	
49.19	Molded part without flow and weld lines in visible or problem areas?			
49.20	Molded part without contamination?			
49.21	Molded part without jetting?			
49.22	Molded part without cold displacements (material)?			
49.23	Injection molding (sprue area) visually OK?			
49.24	Injection molding (sprue) visually OK?			
49.25	Does the removal device function properly?			**Block 7**
49.26	Molded part without diesel effect/burners?			
49.27	Molded part without vinyl record effect?			
49.28	Demolding of the molded part without problems?			
50	**Evaluation of the sampling analysis:**			
50.1	Was the evaluation carried out and documented in full?			
50.2	Are there no defects in the injection mold?			
50.3	Can the basic setting be optimized without negative effects from the mold?			
50.4	Can mold sampling be continued?			

Figure 7.1 Mold sampling checklist: Block 7 (*continued*)

7.1 Producing the First Tooling Samples

Before your sample can be analyzed, it is important that the injection molding process and the injection mold are run in with your basic settings. For this purpose, the set injection molding process must run fully automatically, as only then can the following criteria be set:

- Thermal balancing of the injection mold
- A constant temperature level in the injection molding process
- A constant injection molding cycle

During start-up of the fully automatic injection molding process, the temperature in the mold rises slowly. The reason for this is that the heating effect of the temperature control medium is limited, due to the thermal conductivity of the mold steel and the distances between the temperature control channels and the mold surface. As a result, the run-in time depends on the dimensions of the mold and the mold temperature required for your plastic. In the event of a machine malfunction when the mold has already been in production for some time, it takes 20–30 cycles until thermal balancing is restored and quality parts can be produced again.

7.2 Important Information About Conducting a Mold Sampling Analysis

A large number of practical trials in injection molding companies have revealed considerable deficits in the procedure for finding the necessary optimization requirements for the molded part, the mold and the process. Furthermore, there are difficulties in determining and evaluating the injection molding process with regard to quality and producibility.

The reasons for this are sometimes an unstructured but mostly the lack of a systematic approach adopted by the "mold sampling specialist". What is noticeable is that, when problems arise during sampling, individual measures are tried out on the injection molding machine, without any system behind them. This problem is largely rooted in the vocational training and further training of employees. As a result, a structured approach and method development on the subject of "fault detection, root cause analysis and fault optimization" in the injection molding process should be taught more specifically in the vocational, master craftsmen and technician training schools, and universities.

Sound basic training of employees in a modern company is very important. Particularly in the sampling phase of an injection mold, the "mold sampling specialist" has a great deal of responsibility, as the optimization measures are introduced into the injection mold based on the selected process settings and the subsequent process evaluation. Before we start performing the mold sampling analysis, I would like to explain basic process knowledge and some important content of the mold sampling analysis in more detail, such as

- The injection molding process in general
- The factors influencing the injection molding process and the molded part
- Mapping and evaluating the injection molding process via the cavity pressure
- Thermography in the injection molding process
- The necessary flow rate of the temperature control medium
- The dwell time of the melt in the plasticizing barrel
- Mold spraying, and
- Measurement of the molded parts

7.2.1 The Injection Molding Process

You hear the words "injection molding process" and "process optimization" every day in injection molding companies. But what exactly does the term "process" mean? To answer this question, we will address the following questions in the following sections:

- What is a process in general and how is it defined?
- How is a process structured?
- What characterizes a process?
- Which factors influence a process?

7.2.1.1 Injection Molding Process Definition and Structure

A process in general is the totality of interrelated procedures, operations and activities through which products, energy or information can be transported or transformed. An injection molding process is therefore nothing more than a sequence of logically interrelated activities to create a product.

The injection molding process (Figure 7.2) begins with the input variables (input). These, consisting of the machine setting parameters, are introduced into the process.

The injection molding process is divided into five sub-processes: plasticizing phase, injection phase, switchover to holding pressure, holding pressure phase and cooling phase.

These sub-processes can be considered in the subsequent mold sampling analysis as independent sub-processes with input and output variables. In the injection molding process, there are influencing factors that can have a strong qualitative impact on your process and which are reflected in the output variables as the process progresses.

The output variables are to be considered as the result of your process, as all output variables represent dependent variables of the process. The output variables can be used to express the process characteristics with all its influencing factors. Consequently, output variables are referred to as process parameters. These parameters allow you to analyze and evaluate a process and show the relationships between output and target variables.

Every process has its target variables, which evaluate the entire production process from a technical and economic point of view and therefore form the basis of every optimization. The main objective is to achieve a high level of reproducibility in the injection molding process. In the process, the injection molding machine has the central task of balancing out the external influencing factors. The machine must therefore be able to guarantee reproducibility in the process of pressures, temperatures, speeds or strokes and positions.

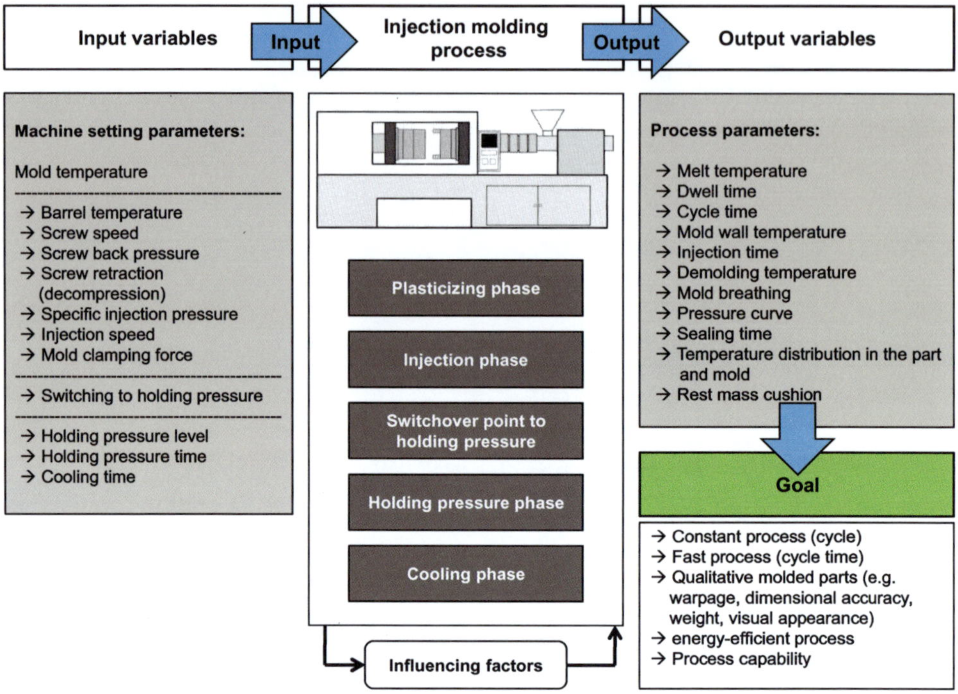

Figure7.2 Structure of an injection molding process

7.2.1.2 Factors Influencing the Injection Molding Process and the Molded Part

The injection molding process is very complex, as many different influencing factors (Figure 7.3), some of which interact with each other, can have an effect on the process and therefore on the quality of the molded part. In order to be able to effectively evaluate an injection molding process, it is first necessary to understand which influencing factors exist and which mechanisms of action the individual influencing factors have on my process or on the quality of my molded part.

Figure 7.3 provides an overview of all factors that can influence the injection molding process and the molded part.

It can be seen here that the input variable "machine setting parameters" and therefore the injection molding machine alone do not provide all the factors influencing the process. Other influencing factors are, for example, the mold temperature control in the injection mold, the plastic properties and the plastic preparation (e. g. pre-drying) as well as the supply of the plastic to the machine or the ambient temperatures of the production areas in the company.

However, the injection molding machine offers the basis for compensating for or improving a large number of influencing factors or disturbance variables thanks to the large number of options for setting the machine parameters. However, this can only be realized if the effects on the process are known. To enable you to better interpret the influencing factors, these are briefly explained.

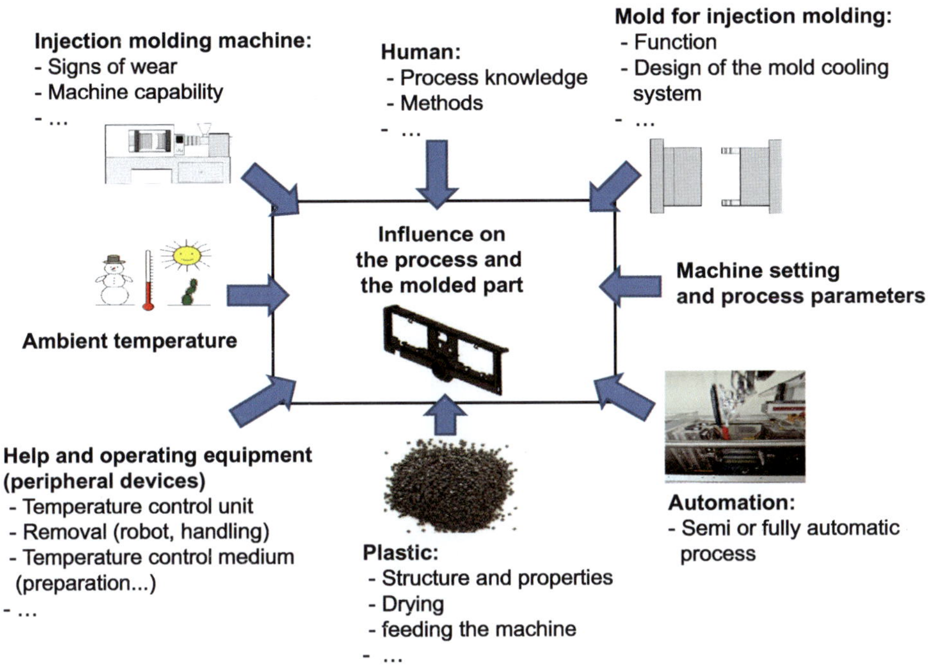

Figure 7.3 Factors influencing the injection molding process and the molded part

Human factor

The “mold sampling specialist” is by rights the first influencing factor mentioned here, as he or she influences the input variables of the process by selecting the machine setting parameters. The two basic pillars that are generally required of the “mold sampling specialist” are a knowledge of plastics and processes and a methodical and systematic approach. This is the basis that a specialist should have. However, it is important that regular further training be conducted internally or externally.

Identification with the company, attention to detail and the personal attitude of the “mold sampling specialist” are crucial. In some cases, there is a lack of awareness that all results, evaluations and further measures, e.g. optimization of the mold, are controlled solely by the specialist. The “mold sampling specialist” is the linchpin of every successful sampling and subsequent series production of the molded parts. A company may use the most elaborate procedural instructions, templates and sampling documentation, but if the motivation and will to get the best out of an injection molding process is not there, these methods will unfortunately prove ineffective and largely worthless.

For this reason, the "mold sampling specialist" must realize how responsible his or her position in the company is, as it is only through his or her work in the company that trouble-free, consistent and economical series production process can take place subsequently.

Influencing factor: injection molding machine

The injection molding machine offers a wide range of possibilities for compensating for or improving the injection molding process, due to the large number of options available via the machine setting parameters. Here it must be ensured that all parameters set by the specialist, e. g. strokes, temperatures and pressures, can also be approached or controlled precisely and consistently in the machine.

In Germany, it is necessary for the manufacturer to test the machine at certain intervals to ensure that it fulfills the German VDMA 24.470 Part 2 standard. Furthermore, the machine setting parameters must be calibrated at the time intervals recommended by the machine manufacturer in accordance with VDMA 24.470 Part 1.

> **NOTE:**
> Find out which standards are prescribed by your customer or recommended by the machine manufacturer in your country.

The following machine setting parameters have a very large influence on the injection molding process if they do not function optimally:

- Hydraulic pressure in the injection barrel (regulated, controlled)
- Injection speed of the screw
- Screw speed
- Barrel temperatures
- Mold clamping force
- Drop in mold clamping force

The VDMA in Germany recommends, for example, calibrating the parameters after 7500 operating hours or once a year.

It must also be ensured that the machine is inspected at regular maintenance intervals for signs of wear and damage to the non-return valve and plasticizing screw. In particular, the leakage flow at the non-return valve and the plasticizing unit should be checked in accordance with VDMA 24.470 Part 1.

NOTE:

Unless the injection molding machine is functioning properly, a clear evaluation of the process in the form of a mold sampling analysis is not possible.

Influencing factor: machine setting and process parameters

In order to better understand this factor's influence on the process and the molded part, let's look at the process from the perspective of the molded part. As is well known, a molded part is created during the injection phase, followed by the holding pressure phase. Clearly, only pressures (e. g. injection and holding pressure) and temperatures (e. g. barrel, melt and mold temperature) and how these change over time influence the molding process for the molded part.

A percentage weighting of the influencing factors pressure and temperature in relation to the molded part and the process is very difficult and sometimes very complex to determine. The reason is that they are often interrelated and sometimes produce different results in terms of quality characteristics.

In order to be able to determine the weighting of the individual factors, design of experiments (*DoE*) is required, which is presented and discussed in more detail in Chapter 8 "Optimization of the basic setting" for the required process optimization.

In principle, however, it can be said that the dominant factors governing the quality of the molded part are the pressure and temperature curve. Perfect process control would exist if the pressure and temperature curve at every point in the mold cavity and consequently the pressure and temperature curve over time were to remain constant from shot to shot. The result would be uniform shrinkage behavior in all parts of the molded part and the absence of internal stresses, all of which would lead to a warp-free molded part. Unfortunately, the reality of injection molding is different, as it is practically impossible to achieve the most constant possible pressure and temperature distribution in the process. The reasons for this are as follows:

- Pressure drop, due to flow resistance during the injection phase, and
- Temperature differences, due to different wall thicknesses and flow paths of the molded part as well as local temperature differences in the mold

As there can be no 100% perfect injection molding process from a physical point of view, the task and motivation of the "mold sampling specialist" should be to approach the ideal state as closely as possible by taking into account all machine setting and process parameters. This means a process setting that achieves pressure and temperature distributions that are as uniform as possible. This is a prerequisite for a constant process and therefore for consistent quality of the molded parts from cycle to cycle.

Influencing factor: plastic

The plastic to be processed also has a major influence on your injection molding process, as appropriate temperatures and pressures must be used to match its properties, for example. The previous chapters have shown you how to set the optimum parameters, such as temperatures and pressures.

A common influencing factor is changes in flow properties (viscosity), due to slight differences in the quality of the individual raw material batches, which can lead to various fluctuations in the process. These fluctuations are clearly reflected in the output variables (process parameters) of the process, as a change in the flow properties results in a change in the injection time and the residual melt cushion.

Material preparation is another factor influencing your process. Here it is important that all hygroscopic plastics, i. e. those that bind moisture from the environment (usually in the form of water vapor from the humidity in the air), must be pre-dried in accordance with the raw material manufacturer's specifications.

If, for example, a hygroscopic plastic enters the injection molding process undried, it will also have a very negative effect on the properties of the plastic and lead to considerable quality problems with the molded part.

PRACTICAL TIP:

Before any sampling, the plastic should be subjected to a residual moisture test and the result compared with the maximum moisture content recommended by the raw material manufacturer.

Influencing factor: injection mold

The injection mold is a fundamental and highly influential factor in your process. Here, the mold can greatly affect the injection molding process and the quality of the molded part as a result of incorrect functioning, such as demolding problems, poor spotting and lack of venting. Furthermore, Section 4.4 "Mold temperature control" has already clearly shown that this is very important for the process, as it influences properties, such as the formation of the crystalline structure, dimensional accuracy, surface quality and the cooling phase of the molded part and thus the cycle time.

For this reason, it is crucial to identify and evaluate all possible sources of error in the mold sampling analysis on the mold in order to then have these corrected by the mold-maker with the help of an optimization loop.

Influencing factor: automation

There is a big difference between a semi-automatic and a fully automatic process. In a semi-automatic process, constant process control cannot be realized, as fluctuations in the cycle occur, due to different removal times, for example, when the molded part is removed manually. Therefore, automatic molded part removal should already take place at this stage of the sampling phase.

PRACTICAL TIP:

In practice, it sometimes happens that the removal system (e. g. handling gripper) is not yet available for initial sampling (FOT) or is still under construction. In this case, the time required to remove the molded part can be specifically estimated and built in as "dead time" in the form of a pause in the process flow/cycle time. This makes it possible, even at this stage, to work with a constant cycle without parts removal.

Influencing factor: ambient temperature

The ambient temperature as an influencing factor is sometimes greatly underestimated or not taken into account at all. However, it plays a decisive role in an energy-efficient process, as draughts from open windows or doors, for example, are responsible for the barrel heater having to constantly heat up or readjust. A constant ambient temperature is also very important, as large temperature differences between summer and winter can cause process fluctuations.

Influencing factor: auxiliary and operating equipment (peripherals)

The last factor influencing the injection molding process is the auxiliary and operating equipment. This includes the temperature control units, cooling systems, removal of the molded parts (robots, handling, etc.), material conveyors and the temperature control medium. All of these peripheral units must be continuously maintained and operate fault-free. If your molded part is not optimally removed, this can have a decisive influence on the cycle time of the process, for example. Furthermore, sub-optimal material feed leads to unnecessary machine downtimes and consequently to an inconsistent process. The temperature control units, cooling systems and the temperature control medium should come in for particular scrutiny. As already described, these peripheral units exert a considerable influence on the quality of the molded part. It must therefore be ensured, on the one hand, that the preparation of the temperature control medium proceeds optimally and, on the other, that the temperature control units have the required capacity to realize a certain flow rate.

7.2.2 Analysis Via the Cavity Pressure Profile

Unfortunately, it is still increasingly common today for specialists to evaluate an injection molding process in the sampling phase using their specialist specialist knowledge alone. Without detailed knowledge of the cavity pressure curve, this method is not optimal and often leads to misinterpretations. Therefore, the general goal of every injection molding company should be the ability to perform process optimization on the basis of the cavity pressure curve.

With the help of the cavity pressure, you can show and evaluate your process visually in order to be able to take the right measures during subsequent optimization (Chapter 8 "Optimizing the basic setting"). With a view to optimizing your work and interpreting the pressure curves, I would first like to answer the following questions about cavity pressure:

- What is cavity pressure in general?
- What does a cavity pressure curve look like?
- What information or knowledge can be gained from the cavity pressure curve?

7.2.2.1 General Information on Cavity Pressure

Cavity pressure is the only physically measurable output variable of the injection molding process that describes all processes during molded part formation in the mold and is directly correlated to the quality (e. g. weight, dimensional accuracy, degree of shaping (contour), sink marks, shrinkage, warpage) and properties (e. g. morphology, internal stresses, crystallinity, orientation) of the molded part. Furthermore, the cavity pressure can be used to visualize a number of factors influencing the molded part formation process in the mold that cannot be precisely controlled or taken into account via the process input variables (input), such as fluctuations in plastic properties and ambient temperatures.

On the basis of the cavity pressure, the following points in the injection molding process can be visualized:

- Optimum injection phase
- Optimum compression phase
- Correct switchover point to holding pressure
- Optimum holding pressure phase

The curve of cavity pressure can be seen as the "fingerprint" of your process, because all the main effects of molded part formation in the mold, including the factors influencing the injection molding process, are mapped here. If the cavity pressure is not optimal or fluctuates, this leads to an unstable injection molding process and consequently to different quality levels of the individual molded parts.

Once an injection molding process has been analyzed and optimized, the aim should be to use the cavity pressure curve as a reference curve (target curve) for monitoring the subsequent series production process. In this way, you can ensure that a deviation in the reference pressure curve can be displayed immediately in the event of fluctuations or changes in the process input variables (input), such as mold and melt temperatures, injection speeds, holding pressure and holding pressure time, plasticizing time and the switchover to holding pressure. This has the great advantage that the required quality requirements can be checked before the molded part is demolded and the molded part can be declared a reject if there is a certain deviation from the reference pressure curve.

7.2.2.2 The Cavity Pressure Curve

The cavity pressure during molded part formation in the mold is divided into three phases (Figure 7.4):

- Injection phase
- Compression phase with switchover to holding pressure
- Holding pressure phase

In each of these three phases, certain machine setting parameters can be used to influence the quality characteristics of your molded part.

	Injection phase	Compression phase	Holding pressure phase
Quality characteristics	Surface (roughness, gloss, color), crystallinity, warpage, orientations in the surface layer area	Molding of the contour, flash formation, weight, dimensional accuracy	Dimensional accuracy, warpage, shrinkage, sink marks and blowholes, internal orientations, stresses, demolding problems
Setting parameters	Injection speed, melt and mold temperature	Switch to holding pressure, melt and mold temperature	Holding pressure, holding pressure time, mold temperature

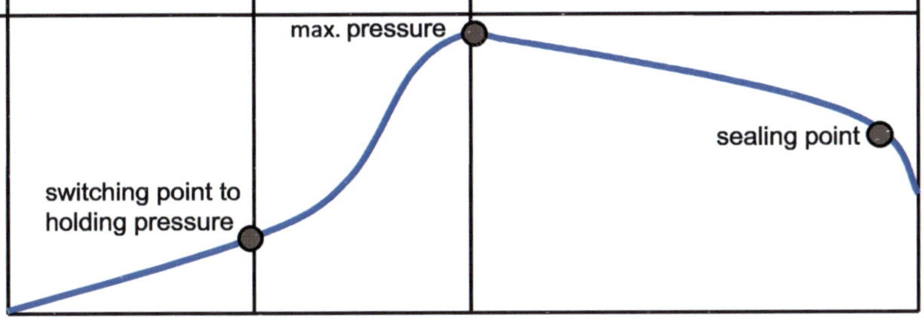

Figure 7.4 The three phases of cavity pressure

The next step is to look at the change in cavity pressure (Figure 7.5) from the beginning of the injection phase to the end of the holding pressure phase. The plasticized plastic melt is injected into the cavity of the injection mold at the start of the injection phase ①. The cavity pressure can only be measured and visualized when the melt front reaches the pressure sensor integrated in the cavity ②. During the injection time, the pressure in the mold increases linearly. As soon as the molded part has been filled 100% volumetrically, the end of the injection phase is reached ③ and the system switches over to holding pressure. The melt is compressed during the compression phase to ensure optimum molding of the molded part. At the end of the compression phase, the maximum cavity pressure ④ is reached and the holding pressure phase follows. If the gate to the molded part has solidified at the end of the holding pressure phase ⑤, the sealing point is reached and the cavity pressure drops to the ambient pressure ⑥.

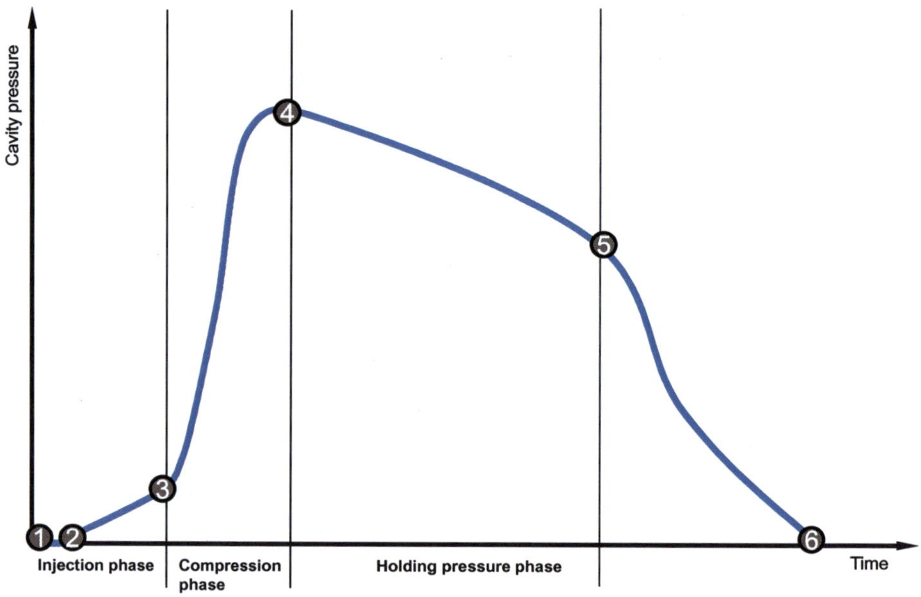

Figure 7.5 The cavity pressure curve

The shape of a cavity pressure curve depends on the following factors:

- Structure of the thermoplastic (amorphous, semi-crystalline)
- Distance of the internal pressure sensor in the mold to the gating of the molded part

Cavity pressure as a function of the structure of the thermoplastic:

Amorphous and semi-crystalline thermoplastics differ in their compressibility and cooling behavior, which leads to different forms of pressure progression in the mold.

In the case of amorphous thermoplastics, the cavity pressure in the holding pressure phase drops in line with the cooling of the molded part, due to increasing viscosity in the molded part. As a result, pressure can no longer be transferred from the screw antechamber of the plasticizing barrel into the mold and the internal pressure drops further to the ambient pressure.

In contrast to amorphous thermoplastics, the pressure in semi-crystalline thermoplastics remains steady at one value for a long period until the crystallite melting point is reached. This is due to the fact that good pressure transfer is initially possible. Once crystallization begins in the molded part, the internal pressure drops steeply to the ambient pressure, due to volume shrinkage.

Figure 7.6 shows the difference between the cavity pressure curve of amorphous and semi-crystalline thermoplastics.

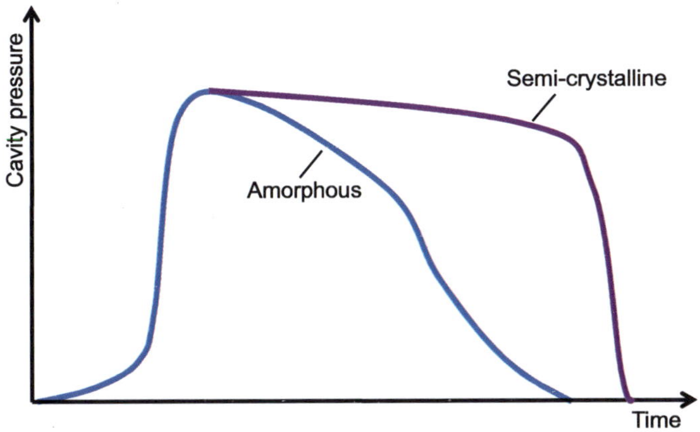

Figure 7.6 Cavity pressure curve of amorphous and semi-crystalline thermoplastics

Cavity pressure depends on the position of the pressure sensor in the mold:

If an internal pressure sensor is installed in your injection mold, it is important to check where it is positioned. In order to analyze and evaluate a process based on the cavity pressure, the sensor should be positioned in the area of the gating (near the sprue), as this is where the melt crystallizes last and a pressure curve can therefore be observed from the beginning of the injection phase to the end of the holding pressure phase. If the pressure sensor is positioned away from the sprue, e.g. at the end of the flow path of your molded part, the shape of the pressure curve is identical, but the pressure level is lower.

The advantages and disadvantages of measuring the internal pressure outside the sprue (e.g. at the end of the flow path) are as follows:

Advantages:

- Pressure-dependent switchover to holding pressure possible
- Mold protection against damage, e.g. to core pulls etc, due to possible pressure peaks
- Avoidance of over-molding in the mold and flash formation on the molded part [1]

Disadvantages:

- Late measurement, due to the longer flow path
- No information about the pressure curve in the injection phase possible
- Pressure measurement incomplete, as the pressure sensor is positioned at the end of the pressure curve in the mold [1]

The ideal case for process analysis and optimization would be a pressure sensor close to the sprue and a pressure sensor remote from the sprue in the injection mold for subsequent process monitoring.

7.2.2.3 Knowledge Gained from a Cavity Pressure Curve

With the help of the pressure curve, it is possible to analyze quality-relevant machine settings, such as holding pressure, holding pressure time and switchover point to holding pressure, based on deviations from the optimum pressure curve (Figure 7.7). Chapter 8 "Optimizing the basic settings" describes how to target the optimum settings.

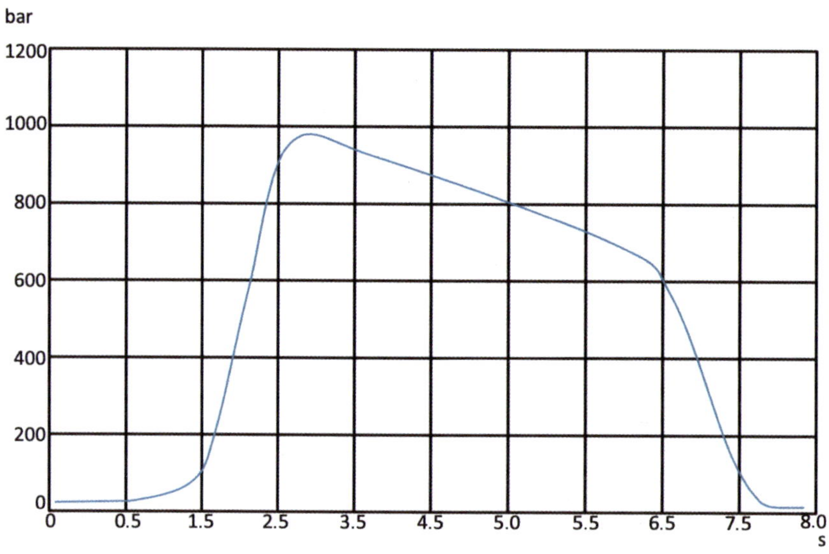

Figure 7.7 Optimum cavity pressure curve

The following subsections explain possible deviations from the optimum course of the cavity pressure (Figure 7.7) and present them in more detail.

Cavity pressure peak:

A cavity pressure peak (Figure 7.8) occurs if the set switchover point to holding pressure occurs too late or the injection speed has been selected too high. The pressure peak leads to fluctuations in dimensional accuracy, internal stresses (over-loaded molded part) and damage to the mold.

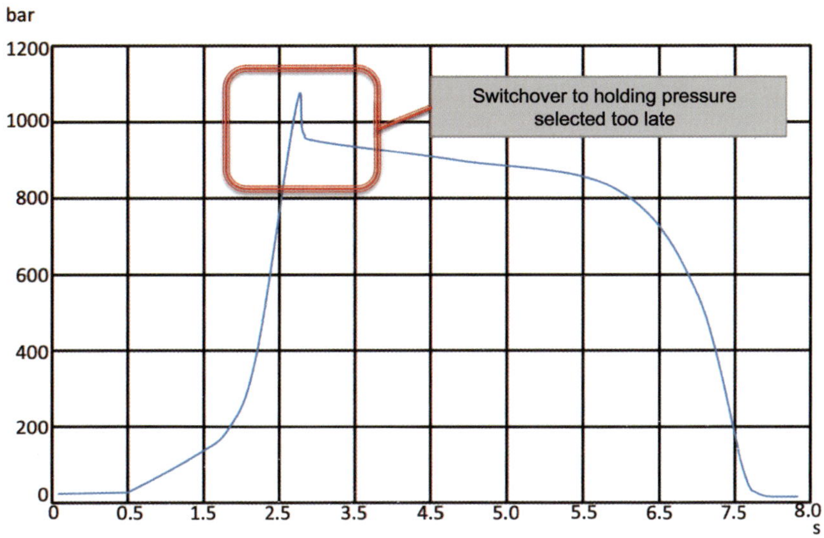

Figure 7.8 Cavity pressure peak

Pressure relief before the decompression phase:

Pressure relief before the compression phase (Figure 7.9) occurs when the switchover point to holding pressure occurs too early, before the molded part has been 100% volumetrically filled. This results in an uncontrollable filling process and a flow delay, which leads to cold displacements. Switching to holding pressure too early also results in molded parts that are not ejected, sink marks, flow marks and a reduction in molded part strength, due to sub-optimal weld lines.

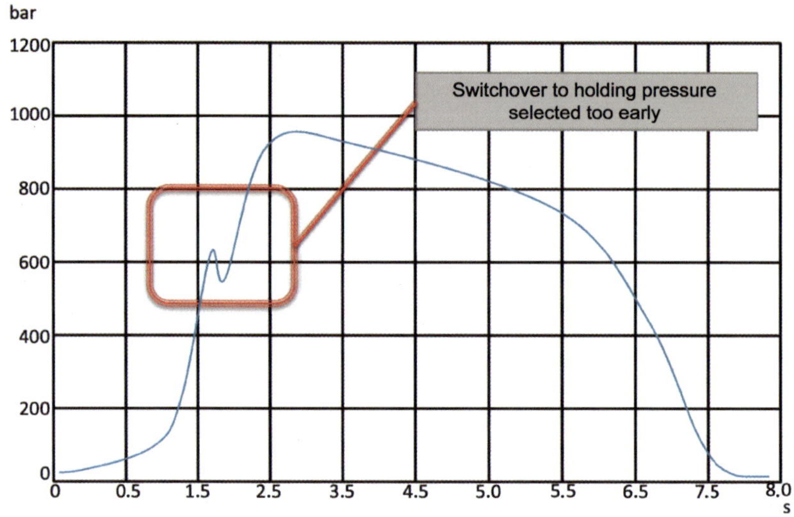

Figure 7.9 Pressure relief before the compression phase

Pressure drop during the holding pressure phase:

A pressure drop during the holding pressure phase (Figure 7.10) shows that the set holding pressure time was too short and consequently the plastic melt can flow back out of the cavity. This means that no volume shrinkage can be compensated for and so sink marks appear on the molded part.

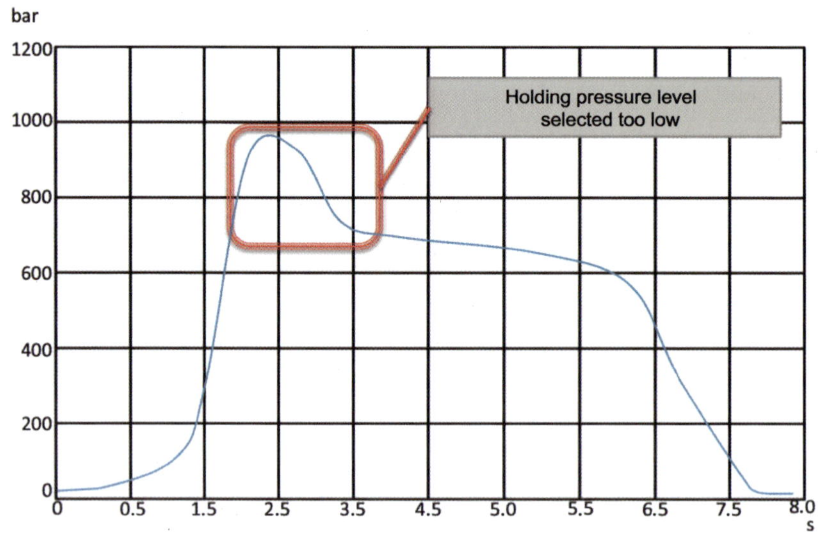

Figure 7.10 Pressure drop during the holding pressure phase

Demolding under residual pressure:

Demolding under residual pressure (Figure 7.11) occurs when the switchover to holding pressure is set too late or the holding pressure is generally too high. The holding pressure should always be set so that the pressure in the mold drops evenly. However, demolding under residual pressure is also possible if the mold gives way, due to the high pressure, and deforms or the machine opens. In addition to the high stresses in the molded part, both cases also pose a high risk to the mold, which can be damaged quickly under these circumstances.

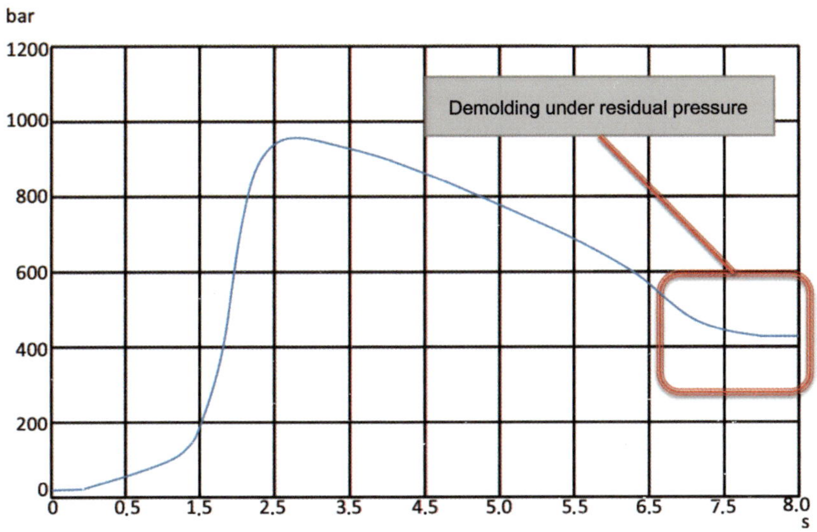

Figure 7.11 Demolding under residual pressure

7.2.3 Mold Sampling Analysis Using Thermography

Infrared thermography involves measuring the surface temperature of an object with a thermographic camera (thermal imaging camera). In the injection molding sector, measurements can be taken directly on the molded part. They enable the visualization of hot spots and the detection of temperature differences, and the cooling behavior of a molded part can also be observed in this way. The "mold sampling specialist" can use this information to assess whether the process is progressing optimally or which optimizations might be necessary during sampling. In many cases, problems occur during the sampling phase, such as:

- Cycle time longer than calculated in advance
- Strongly fluctuating molded part quality

- Flash formation
- Sink marks
- Dimensional accuracy out of tolerance
- Different shrinkage potentials and warpage

These problems are often caused by an inhomogeneous heat budget in the mold, such that up to approx. 60% of all molded part defects that occur in practice are due to sub-optimal temperature control of the mold. Therefore, in addition to cavity pressure, a thermal analysis by means of thermography is very important for detecting thermal sources of faults in the mold and on the molded part at an early stage.

Despite these advantages, the use of thermography in mold sampling is very rare in injection molding companies. As a result, it is not possible to draw optimal conclusions about thermal sources of faults during mold sampling, as shown in Table 7.1.

Table 7.1 Thermal Causes of Faults [2]

Appearance	Thermal problem	Most common cause
Dimensional problems, lack of mechanical properties	Excessive temperature deviations on the mold wall	Asymmetrical temperature control channel layout, inadequate heat dissipation, bridging of temperature control channels
Molded part warpage	Excessive temperature deviations on the mold wall, partially or on the entire molded part	
Surface markings, due to gloss and matt spots, flash formation	Thermal molding of inserts and ejector pins, domes, ribbing and openings	Inadequate heat dissipation, inadequate insulation of hot runner systems and hot runner nozzles, non-temperature-controlled sections of molded parts
Cooling times/cycle times significantly too long	Inadequately designed mold temperature control, high pressure losses within the temperature control system, hot spots on the molded part	Blocked cooling channels, poor technical condition of the temperature control and cooling units, aggressive water condition, untreated or inadequately treated water

Before I go into further detail about process analysis by means of thermography, the "Background knowledge" information box below explains more about the topic of thermography and its use in injection molding.

BACKGROUND KNOWLEDGE:

Use of infrared measurement technology in injection molding

What is thermography?

Thermography is the use of a camera to visualize and measure the thermal energy (heat energy) emitted by an object. Thermal or infrared energy is light whose wavelength is too long to be perceived by the human eye. This energy is the part of the electromagnetic spectrum that we perceive as heat. In contrast to the case for visible light–namely that everything which our eyes are able to see must necessarily be illuminated, whether by the sun or other light sources—in the infrared world, every object whose temperature is above absolute zero radiates heat.

Therefore, even very cold objects, such as ice cubes, emit infrared radiation. The higher the temperature of an object, the more intense is the infrared radiation it emits. We can detect things in the infrared range that are otherwise invisible to our eyes. Powerful infrared camera systems can detect and measure this radiation at temperatures below –40 °C. Infrared cameras generate images of the invisible infrared or thermal radiation and thus enable precise non-contact temperature measurements in the range –40 to over 2000 °C. The thermal infrared spectrum (Figure 7.12) begins directly after visible light at approx. 0.7 mm and extends up to a wavelength of 14 mm [3].

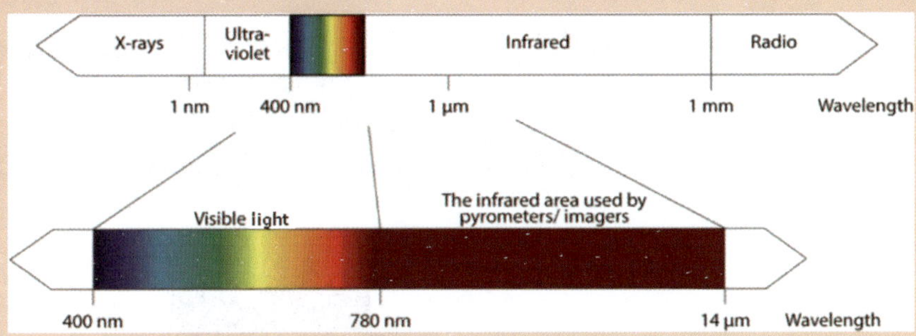

Figure 7.12 Thermal infrared spectrum (image source: Optris GmbH)

What requirements should an optimal system for use in the plastics industry meet?

First of all, the wavelength is important. As the majority of measurements in the injection molding process are taken directly on the molded part and high emission gradients are generally used here, a camera in the long-wave spectral range (7 to 13 µm) should be chosen. Some plastics are "transmissive" in the short-wave range (3 to 5 µm) and do not necessarily allow a clear determination of the surface temperature here.

Another important factor is the geometric resolution. The parts to be measured or “hot spots” are sometimes quite small and place high demands on the geometric resolution (number of pixels) of the camera. Rarely are 160 × 120 cameras found in this area. Very popular are 320 × 240 camera systems. In the future, 640 × 480 systems will prevail.

The temporal resolution of the camera systems used is just as important. In injection molding, it is very important to create images before and after changes in mold temperature control. It is important to select a defined point in time after mold opening or after removal in order to be able to document the resulting changes to the molded part at the same time. Functions such as slow image build-up or averaging algorithms are anything but helpful here.

Another important point is remote operability of such a system. Many users mount the camera on the machine in order to have as clear a view as possible of the molded part that has not yet been demolded and to be able to create the resulting temperature distributions directly after mold opening (triggered). Hand-held remote controls with a screen or systems with real-time data output directly via a connected laptop are advantageous for this purpose. A digital photo system integrated into the infrared camera system has proven to be very helpful in practice. This can be used to document visible surface impurities in addition to the thermal image.

Many users choose the option of recording dynamic data and subsequently documenting temperature curves over time, as the heat capacity factor must not be underestimated for molded parts and certain areas in the demolded parts only reach their actual maximum temperature some time after removal. Other important factors when selecting an infrared system for injection molding are:

- Absolute accuracy
- High reproducibility
- Detector resolution (number of pixels)
- Spot size (distance to the measurement object)
- Thermal sensitivity
- Powerful software packages [2]

What points must be paid particular attention to when taking thermal images?

The accuracy achieved in the measurements is ±1 °C. A prerequisite for this is correct setting of the emissivity of the object to be measured. Working with infrared technology requires extensive knowledge of how to handle the system and the physics of the basic mechanisms of the emission, transmission and reflection of radiation. However, if these three effects are correctly assigned and taken into account, the “mold sampling specialist” is able to analyze and evaluate the molded parts for optimum condition. To do this, the specialist needs the following information:

What is meant by emission, transmission and reflection?

Firstly, the radiation of the object detected by the thermal imaging camera consists of emission, transmission and reflection (Figure 7.13). The sum of these three variables corresponds to 100% or is set to 1. The intensity of the emitted radiation depends on the emissivity of the material.

- The *emissivity* (ε) is the ability of a material to emit (emit) infrared radiation
- The *reflectance* (*R*) is the ability of a material to reflect infrared radiation. This depends on the surface properties and the type of material
- The *transmittance* (*T*) is the ability of a material to transmit infrared radiation. This depends on the thickness and type of material and indicates the material's permeability to IR radiation

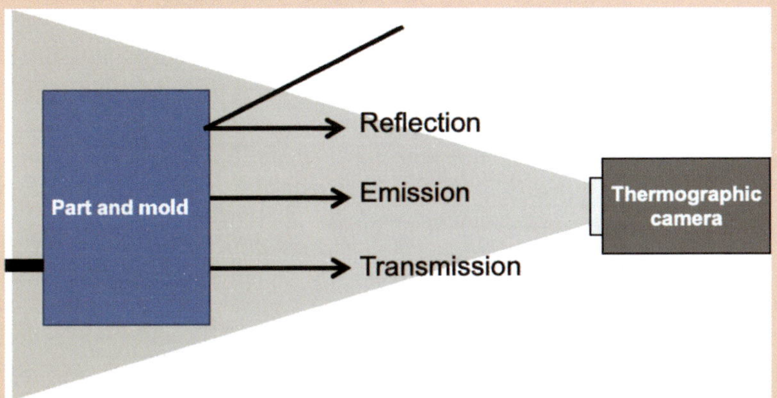

Figure 7.13 Radiation detected by a thermographic camera (schematic)

These three variables can assume values between 0 and 1 (or between 0 and 100%).

To ensure smooth measurement using thermography, the following points must be observed in practice during mold sampling:

1. Set the emissivity correctly:

- The molded parts are usually measured in the temperature range from +20 °C to +300 °C. Here, the emissivity of almost all plastics is between 0.80 and 0.95
- The greater the difference between the object temperature and the ambient temperature and the lower the emissivity, the greater the error if the emissivity is incorrect
- At temperatures greater than the ambient temperature:
  - Emissivity values that are set too high result in temperature displays that are too low
  - Emissivity settings that are too low result in temperature displays that are too high
- At temperatures lower than the ambient temperature:
  - Emissivity values that are set too high result in excessively high temperature displays
  - Emissivity settings that are too low result in temperature displays that are too low

2. Ensure the correct distance between the object and the camera:

For this purpose, the ratio of measuring distance to measuring spot must be known from measuring spot diagrams. You can obtain this information from the manufacturer of your camera

3. Remove possible interfereing factors, such as ceiling lamps and radiant heaters:

If interfering factors (e. g. radiators, lamps, refrigeration units, etc.) reflect on the surface of the object, this will significantly distort the measurement result. To avoid reflections from the surroundings, shielding with a cardboard box, for example, is very useful

7.2.3.1 Thermography on Molded Parts

The most common type of IR thermography is imaging of the molded parts directly after removal from the injection mold or, in the case of larger articles, during removal. Figure 7.14 shows infrared thermographs with corner and rib areas where the temperature control is not optimal, as well as asymmetrical temperature control and hot spots.

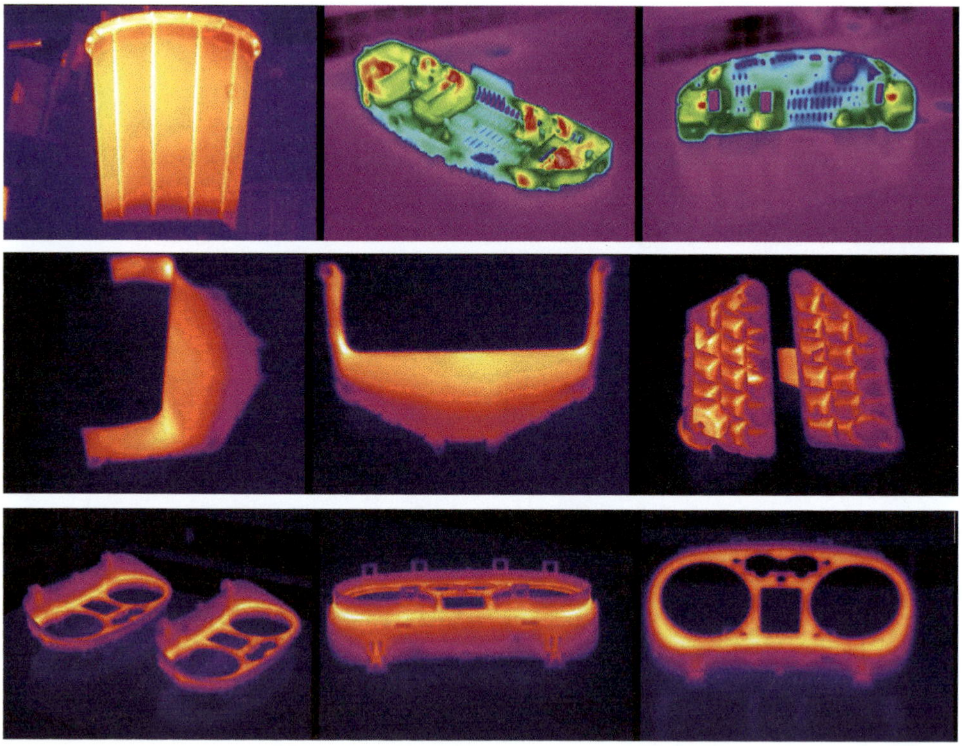

Figure 7.14 Examples of IR images of various molded parts (image source: GTT Willi Steinko GmbH)

7.2.3.2 Thermography on Injection Molds

In principle, injection molds (Figure 7.15) cannot be thermographed without gross errors or, if so, only to a limited extent. Reflections and radiation errors caused by high-gloss surfaces mean that the surface temperature of the injection mold cannot be measured with sufficient accuracy. Only blackening of the mold enables an exact measurement of the mold surface. The blackening method cannot be used in injection molding. However, experienced thermographers can carry out measurements on metals using appropriate aids such as paint, spray or special films that allow of very precise conclusions about the possible variation in temperature of the mold surface.

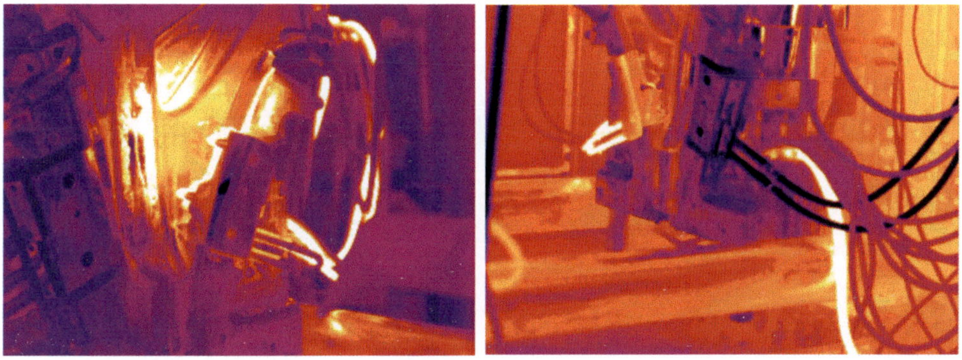

Figure 7.15 Example of mold thermographs (image source: GTT Willi Steinko GmbH)

7.2.3.3 Thermography on Hot Runner Systems

Hot runner systems that are not optimally designed or not sufficiently thermally insulated are largely responsible for an inhomogeneous heat budget within the mold. It is not possible to precisely assign this to incorrect design or to inadequate insulation of the system when the mold is installed. The mold must be dismantled and the hot runner checked in order to be able to make an informed statement in the event of possible problems during sampling. In practice, thermographs of molded parts often reveal heat pockets (Figure 7.16), which cannot be assigned with certainty.

After checking the temperature control channel layout, it may ultimately be ruled out that the temperature control is responsible, but the cause or origin of these heat pockets will not be sufficiently clear.

In order to identify the source of the thermal fault by means of thermography, sequential thermography, i. e. thermal images of the heating behavior of the hot runner system taken at defined time intervals, should be examined. The sequential thermal images (Figure 7.17) reveal asymmetrical heating behavior by the hot runner system, which leads to an uneven temperature distribution within the mold and, in the event of inadequate mold temperature control, is also visible on the molded part as a heat pocket.

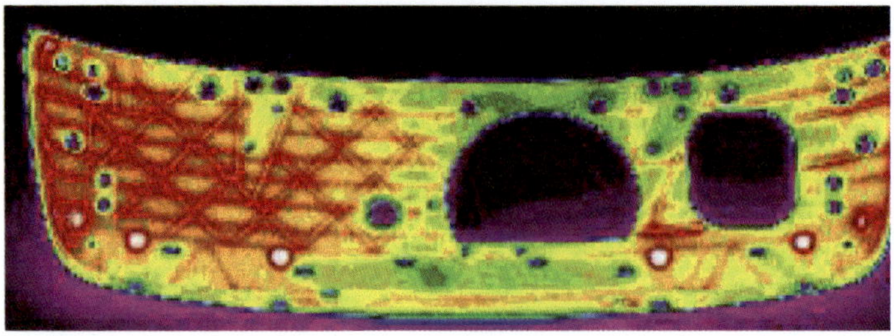

Figure 7.16 Example of a heat pocket (molded part with equal wall thicknesses) (image source: GTT Willi Steinko GmbH)

The temperature control of the manifold bars and nozzles is particularly critical in multiple hot runner systems. In practice, mold sampling often reveals incorrectly functioning and insulated systems, which can only be clearly identified by means of thermography and allow the problems to be targeted and resolved.

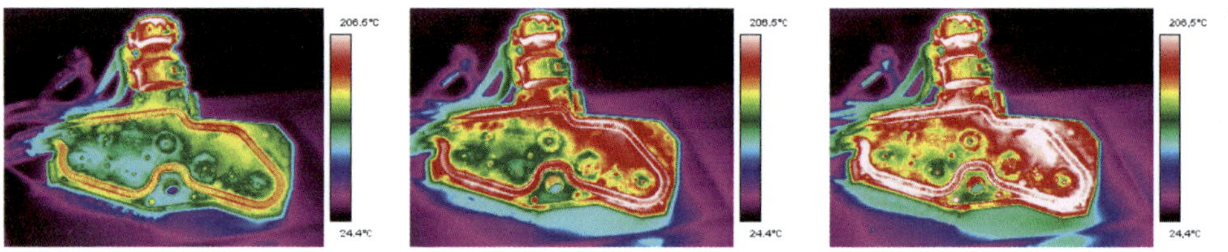

Figure 7.17 Example of sequential thermography of the heating behavior of a hot runner (image source: GTT Willi Steinko GmbH)

7.2.4 Analyzing the Required Flow Rate of the Temperature Control Medium

As already described in Section 4.4.3 "Checking the flow rate of the temperature control medium", the flow rate (l/min) of the temperature control medium has a direct influence on the heat transfer in the temperature control channel of the injection mold and can therefore contribute to the causes of thermal faults or sources of interference in an injection mold.

Most of the thermal energy (enthalpy) of the plastic melt generated via the barrel temperature and friction must be dissipated again via the temperature control medium (e. g. water). It is therefore important to check the amount of heat that needs to be introduced into the mold during the production of your molded part and dissipated again via the temperature control medium. To do this, the amount of heat (Q) and the required flow rate (l/min) must be calculated. The specific enthalpy difference (kJ/kg) between the measured melt temperature (T_M) and the required demolding temperature (T_E) of the molded part plays a decisive role here. This enthalpy difference is different for each plastic and can be taken from enthalpy diagrams (Figure 7.18 and Figure 7.19).

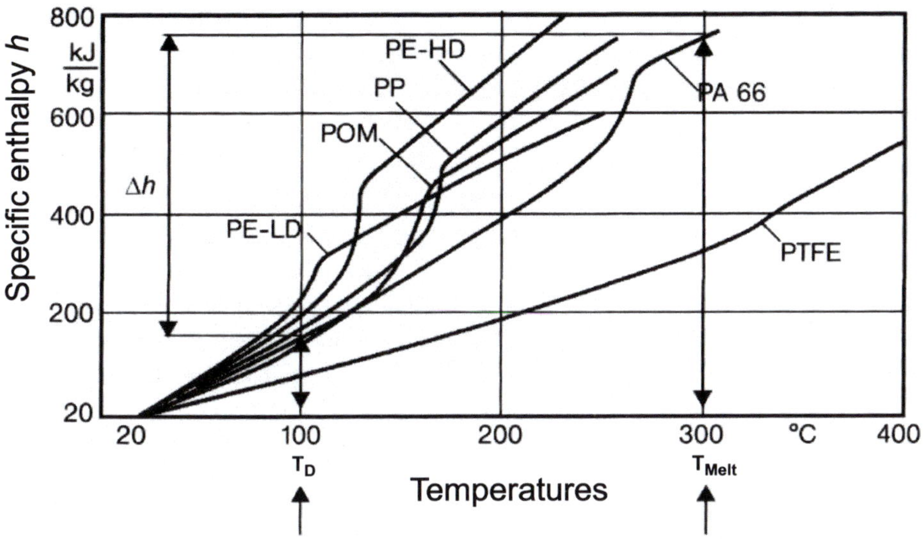

Figure 7.18 Specific enthalpy diagram using the example of semi-crystalline thermoplastics (source: Optimization of injection molding processes, Willi Steinko, 2007 Carl Hanser Verlag, Munich, page 276, Figure 7.74)

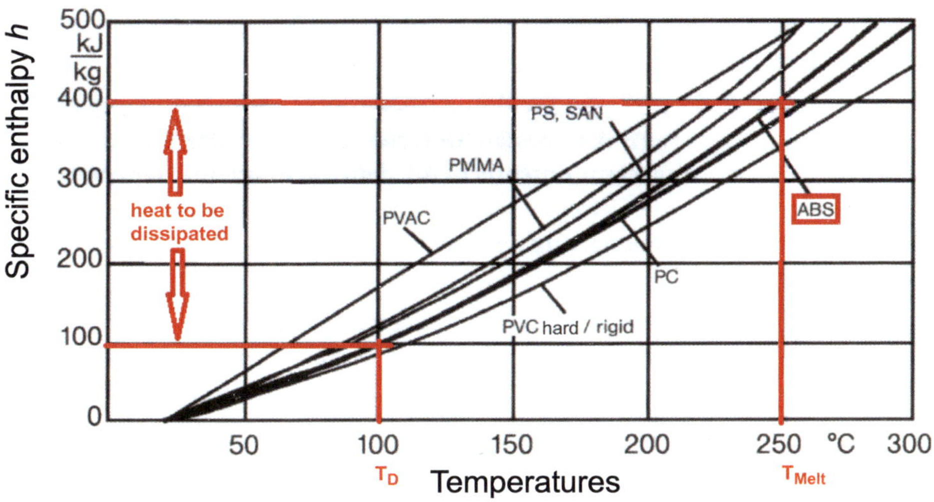

Figure 7.19 Specific enthalpy diagram for the example of amorphous thermoplastics (source: Optimization of injection molding processes, Willi Steinko, 2007 Carl Hanser Verlag, Munich, page 276, Figure 7.75)

For optimum functioning of the temperature control system designed for the mold, a temperature difference of 2–3 °C between the flow and return temperature of each temperature control circuit to the temperature control unit should not be exceeded in practice. Higher temperature differences indicate an uneven temperature distribution in the mold and should be examined more closely using thermographic images of the injection mold.

With the help of the following practical example, I would now like to show you how you can easily calculate the required heat quantity Q and the required flow rate in l/min of the temperature control medium for your molded part without a great deal of mathematical knowledge.

PRACTICAL EXAMPLE:

Given:

- Molded part weight: 161.17g = 0.16 kg
- Specific enthalpy (Δh) according to Figure 7.19: 300 kJ/kg
- Melt temperature (T_M): 250 °C
- Demolding temperature (T_D): 100 °C
- m_p: Cycle time: 40 s → 1.5 molded parts per minute (productivity) → 40 parts/h * 0.16 kg = 6.4 kg/h material
- ΔT = permissible temperature increase between flow and return selected at 2 K
- Safety factor (S) with 1.5
- Specific heat capacity of water (c_W): 4.19 kJ/kg/K
- Density of the temperature control medium water (ρ_W) = 1 kg/dm^3
- Q_{HK}= Installed heating capacity (kW)
- Q_{HR}= Loss factor of the hot runner (0.4 to 0.6)
- **Sought:**
- Total amount of heat to be dissipated Q (kW)
- Flow rate of the temperature control medium (l/min)

Solution:

1. Calculation of the cooling capacity:

$$Q = \left(\frac{m_p \times \Delta h \times S}{3600} \right) + \dot{Q}_{HS} \times \dot{V}_{HS}$$

Substituting:

$$Q = \left(\frac{6.4 \,\frac{\text{kg}}{\text{h}} \times 300 \,\frac{\text{kJ}}{\text{kg}} \times 1.5}{3600} \right) = 0.8 \text{ kW}$$

(Note: Example does not include hot runner mold)

2. Calculation of the required flow rate (l/min):

$$V = \frac{Q \times 60}{c_W \times \Delta T \times \varrho_W}$$

Substituting:

$$V = \frac{0.8\,\text{kW} \times 60}{4.19\,\frac{\text{kJ}}{\text{kgK}} \times 2\,\text{K} \times 1\,\frac{\text{kg}}{\text{dm}^3}} = 5.72\,\text{l/min}$$

This example calculation shows that a heat quantity of approx. 0.8 kW must be dissipated per molded part. To maintain a maximum temperature increase of 2 K, this corresponds to a required flow rate per temperature control circuit of approx. 5.72 l/min.

After you have checked the flow rate (l/min) of the new mold in Section 4.4.3 "Checking the flow rate of the temperature control medium", you can compare and evaluate the practical value with the current theoretically required flow rate (l/min) under process conditions.

7.2.5 Analyzing the Dwell Time of the Melt in the Barrel

The dwell time $t_{\text{dwell time}}$ is the time that a plastic pellet requires for transport, from being drawn into the barrel to being injected into the mold. If a mold with a hot runner system is used, this dwell time must be added to the dwell time in the barrel.

In most cases, only moderate attention is paid to the dwell time in practice, although it should deserve more attention, due to its impact. If the plastic melt remains in the barrel of the injection molding machine for too long, this can result in signs of degradation of the plastic in the form of optical discoloration on the molded part. Another consequence of an excessively long dwell time is the degradation of the required mechanical properties of the molded part.

In general, however, it must be assumed that material damage already occurs before optical phenomena indicate this. Therefore, the dwell time in the injection molding process must be examined during the mold sampling analysis.

The recommended dwell time for your plastic melt can be found in the material processing data sheets of the raw material manufacturer. If no data sheets are available, you can also request them from the raw material manufacturer's application technology department.

PRACTICAL TIP:

A guide value for the dwell time from practical experience is between five and eight minutes in the injection molding process. Dwell times between three and five minutes and between eight and ten minutes are to be considered critical and must be checked in the mold sampling analysis.

7.2.5.1 Determining the Average Dwell Time ($t_{\text{dwell time}}$)

To be able to check your dwell time, I would like to present two methods:

- Empirical determination of the dwell time via a coloring test
- Mathematical determination of the average dwell time

Empirical determination of the average dwell time via a coloring test

Empirical determination of the dwell time via a coloring test is only suitable if you need to color your plastic, as you are dealing with a natural material and a matched color batch. If you do not have a colored molded part, the barrel or hot runner system will require more cleaning following this test. This method is less suitable for transparent or optical molded parts.

Procedure:

- **Step 1:** Allow the plasticizing screw to run empty
- **Step 2:** As soon as the screw flights become visible, add a small amount of color batch at the end of the set plasticizing stroke (screw stopped)
- **Step 3:** Once the color batch is in the feed area at the start of transport, the material to be processed in your molded part can now be conveyed back into the material hopper
- **Step 4:** The cycle count begins with the subsequent injection process
- **Step 5:** The subsequent molded parts must now be visually assessed for the color intensity of the color batch. If the molded part with the highest color maximum is present, this molded part corresponds to the average dwell time

$$\text{Average dwell time } (t_{\text{dwell time}}) = \text{Number of cycles} \times \text{Cycle time (s)} \tag{7.1}$$

Mathematical determination of the average dwell time

With this method, you can calculate the average residence time t_V very quickly, regardless of the plastic being processed (Equation 7.2).

$$t_{\text{dwell time}} = \frac{\text{Max. machine metering volume } (\text{cm}^3) \times \text{Cycle time (s)}}{\text{Required shot volume } (\text{cm}^3) \times 60} \tag{7.2}$$

7.2.6 Checking the Existing Material Dryer Volume

In this section of the mold sampling analysis, you must check whether the available material dryer has the necessary capacity for processing pre-dried plastics.

If the capacity of the dryer is too small, the moisture cannot be properly removed from the material when the process is running. The consequences are changes in the viscosity of the plastic melt, impairment of the mechanical properties and surface defects in the form of streaks or bubble formation.

If, on the other hand, the volume of the barrel is too large, there is a risk of overdrying the material. The consequences can include changes in the viscosity of the plastic melt and degradation (thermal damage) of the molecular chains.

Using a practical example, I would like to show you how you can check your dryer volume quickly on site while the process is running.

PRACTICAL EXAMPLE:

Given:

- Molded part: PA 6.6
- Bulk density: 0.67 kg/dm³
- Shot weight: 80 g
- Required drying time: 4 h

Procedure in sub-steps:

Step 1: Determining the bulk density (ρ) of the plastic pellets to be processed

$$\rho_{\text{bulk density}} = \frac{\text{Weight (m)}}{\text{Volume (V)}} = \left(\frac{\text{kg}}{\text{dm}^3}\right)$$

→ Weigh out 1 dm³ (1 liter) of plastic pellets

Step 2: How much material is needed per hour?

$$\text{Material per hour} = m_{\text{shot weight}} \times \frac{3600\ \text{s/h}}{t_{cycle} \times 1000\ \text{g/kg}}$$

$$\text{Material per hour} = 80\ \text{g} \times \frac{3600\frac{\text{s}}{\text{h}}}{35\ \text{s} \times 1000\frac{\text{g}}{\text{kg}}} = 8.22\frac{\text{kg}}{\text{h}}$$

Step 3: How much material is required in the dryer?

$$m_{\text{material in the dryer}} = m_{\text{material requirements per hour}} \times t_{\text{required drying time}}$$

$$m_{\text{material in the dryer}} = 8.22\frac{\text{kg}}{\text{h}} \times 4\text{h} = 32.91\ \text{kg}$$

Step 4: What volume is required for the current process cycle?

$$\text{Required dryer volume} = \frac{M_{\text{amount of material in the dryer}}}{\rho_{\text{bulk density of the plastic pellets}}}$$

$$\text{Required dryer volume} = \frac{32.91\ \text{kg}}{0.67\frac{\text{kg}}{\text{dm}^3}} = 49.12\text{dm}^3 \approx 50\ \text{liters}$$

Answer:

According to the values given above, a dryer capacity of 50 liters is required for the current running process.

7.2.7 Checking the Touching Surfaces in the Mold

Touching (adjustment of the contact surfaces) of the injection mold should already have been carried out by your mold maker with the aid of a spotting press. This enables the mold maker to fine-tune the mold halves to each other. A lack of spotting may be responsible for possible faults in the molded part. Possible faults are flash formation, due to insufficient spotting in the parting line, burns (diesel effect) on the molded part due to excessive surface pressure caused by air in inserted vents that cannot escape. One possible cause is that the mold has not been spotted under process conditions in which temperatures and pressures act as a function of time. It is therefore always advisable to create a spotting image as a check during the mold inspection, or to evaluate it and archive it in the form of a photo for later mold documentation.

7.2.8 Checking the Dimensional Accuracy of the Molded Part

In most cases, dimensional accuracy of the molded part plays a very important role, especially when it comes to technical or functional molded parts. Therefore, before starting the mold sampling analysis of the basic setting, it is important to check whether the current actual dimensions of the molded part are within the agreed permissible tolerances of the nominal dimensions specified in the drawing.

Depending on the company or the complexity of the molded part, the dimensional accuracy is usually checked by the "quality assurance/metrology specialist" using conventional measuring methods, such as calipers, micrometers, height gauges, or with a coordinate measuring machine or with the help of a 3D scan. It is crucial here that the "mold sampling specialist" receives a measurement report showing all target and actual dimensions of the molded part. This enables the "mold sampling specialist" to identify the cause of any deviations during the mold sampling analysis of the basic setting. Furthermore, the measurement report helps the specialist on the injection molding machine to achieve improvements in the subsequent injection tests (Chapter 8 "Optimizing the basic setting").

7.3 Starting the Mold Sampling Analysis of the Basic Setting

The quality, appearance, mechanical properties and functionality of your molded part are influenced by numerous factors, including the design of the molded part, the functionality of the injection mold and the correct setting of the machine parameters.

The previous chapters of this book have described in detail the necessary physical and process engineering principles for an optimum approach to the basic setting of your injection molding process. Experience shows that the molded parts that are produced with a basic setting do not yet meet the quality requirements and considerable process fluctuations can occur. Consequently, it is now important to examine your basic setting for all possible sources of faults or malfunctions, so that these can be specifically eliminated during optimization.

To this end, I would like to introduce you to a systematic, structured method for discovering all possible sources of faults and malfunctions on the molded part, mold and injection molding process and assigning them to specific causes.

The focus of the mold sampling analysis must be on the quality of the molded part, the producibility of the mold and the overall need to optimize the injection molding process and the injection mold.

Only in this way can the right decisions be made in a targeted and effective manner when determining the optimization measures or optimization strategies.

The mold sampling analysis (Figure 7.20) begins with the visual and thermal evaluation of the molded part, as any faults that occur here can be attributed to the injection molding process or the injection mold.

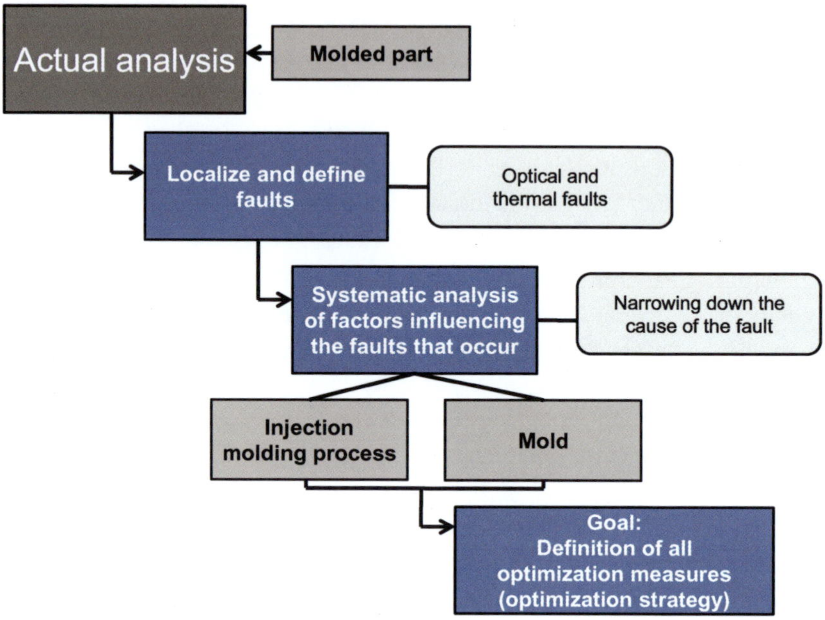

Figure 7.20 Schematic flow of the mold analysis

PRACTICAL TIP:

Before the mold sampling analysis can begin, the injection molding process must be run in fully automatic mode and the mold must be in thermal equilibrium (see Section 7.1 "Producing the first tooling samples").

7.3.1 Implementing a Mold Sampling Analysis

In the following sections, I would like to show you how to carry out the mold sampling analysis in sub-steps.

Step 1: Localize and define faults

For effective troubleshooting during mold sampling, it is crucial to locate and define the faults that occur on the molded part and then analyze the cause. Localizing and defining means first identifying the location of the fault and then assigning it to a known fault pattern, e. g. sink mark.

Unfortunately, in injection molding there is usually not just one cause responsible for a fault pattern. Several causes (Figure 7.21) may be responsible and they sometimes interact with each other. In addition to the cause of the fault, the time at which it occurred should always be narrowed down more precisely.

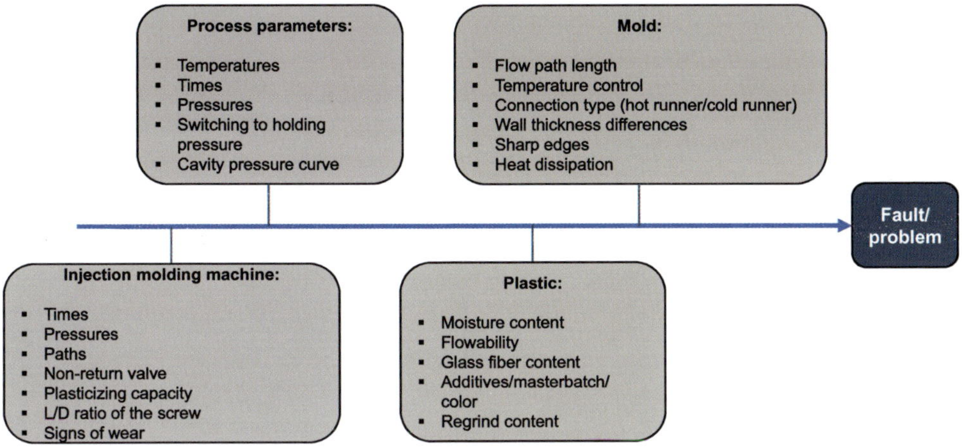

Figure 7.21 Possible causes for the occurrence of a fault pattern

The molded part is first inspected visually and then with the help of thermal images (thermography) for thermal faults (Table 7.2). If faults are found, these must be defined and assigned to a known fault pattern (e.g. sink marks or hot spots).

Table 7.2 Localize and Define Faults on the Molded Part

Optical faults on the molded part:	Thermal faults on the molded part:
Color streaks	Inadequate temperature control
Burn streaks	Hot spots on the molded part
Glass fiber streaks	Asymmetrical cooling behavior
Moisture streaks	Demolding temperature too high
Sink marks	Heat pockets
Gloss, differences in gloss	
Unmelted material	
Weld lines, flow lines	
Jetting	
Diesel effect, burns	
Air hooks, air streaks	
Visible ejector marks	
Thread formation	

Table 7.2 Localize and Define Faults on the Molded Part (*continued*)

Optical faults on the molded part:	Thermal faults on the molded part:
Vinyl record effect	
Dark spots	
Dull spots in the gate area	
Flaking of the surface layer	
Cold displacements	
Gray/black cloud formation	
Incompletely filled parts	
Molded part over-molded (flash formation)	
Deformation during demolding	
Warpage of the molded part	
Stress cracks, white cracks on the molded part	
Air inclusions, bubble formation	
Blowholes	

Step 2: Localize the cause of the fault:

Once you have localized and defined all optical and thermal faults that occur, these must be assigned to their possible causes. This is the only way to rule out a number of possible causes of the fault at this stage. Table 7.3 presents a guide to the most common faults found in a sampling; the faults in the molded part are narrowed down to either the “injection molding process” or the “injection mold”.

Table 7.3 Frequently Occurring Faults during Mold Sampling

Faults on the molded part:	Cause of the fault											
	Process				Mold							
	Plasticizing	Injection phase	Holding pressure phase	Cooling phase	Demolding	Gating design	Temperature	Mold venting	Touching parts	Hot runner system	Stability	Mold design
Optical faults:												
Color streaks	X	X				X						
Burn streaks	X	X								X		
Glass fiber streaks	X	X	X				X					
Sink marks			X			X	X					
Gloss, differences in gloss	X	X	X			X	X	X				
Unmelted material in the molded part	X											
Weld lines, flow lines	X	X	X			X	X	X				
Jetting		X				X						
Diesel effect, burns		X						X	X			
Air hooks/air streaks	X	X										
Visible ejector marks			X	X	X							
Thread formation	X			X		X				X		
Plate-out	X	X						X				
Vinyl record effect	X	X					X					
Dark spots	X									X		
Dull spots in the sprue area		X				X						
Flaking of the surface layer	X	X					X					
Cold slugs, cold displacements	X	X				X						

Table 7.3 Frequently Occurring Faults during Mold Sampling (*continued*)

Faults on the molded part:	Cause of the fault											
	Process				Mold							
	Plasticizing	Injection phase	Holding pressure phase	Cooling phase	Demolding	Gating design	Temperature	Mold venting	Touching parts	Hot runner system	Stability	Mold design
Gray/black cloud formation	X											
Incompletely filled parts	X	X	X			X	X	X				
Molded part over-molded (flash)	X	X	X						X		X	
Warpage of the molded part			X	X	X							
Distortion of the molded part	X	X	X	X	X	X	X					X
Stress cracks, white crack on the molded part			X	X	X		X				X	
Air inclusions, bubble formation	X	X						X	X			
Blowholes			X			X						
Thermal faults:												
Inadequate temperature control at molded part corners, ribs, domes and cores							X					X
Hot spots on the molded part										X		X
Heat pockets on the molded part										X		X
Demolding temperature too high				X								

Step 3: Targeted analysis of the injection molding process and injection mold:

Once you have localized and defined the faults occurring on the molded part and assigned them either to the injection molding process or to the injection mold, targeted analysis of the possible causes or sources of the faults can begin.

1. *Identification of possible influencing factors:*

   To help you with this, use Table 7.4 to find the possible influencing factors of the machine setting parameters and the necessary checks of the process parameters for the faults occurring on the molded part. Similarly, use Table 7.5 to find the possible influencing variables that need to be checked on the injection mold

NOTE:

In Table 7.4 , "X" stands for machine setting parameters that can possibly have an influencing effect on your molded part fault. The arrows also indicate whether these parameters should be increased (↑) or decreased (↓). All other fields in Table 7.4 and Table 7.5, that are marked with a box (□) must be specifically checked and evaluated in the mold sampling analysis

2. *Make a summary of the analyzed influencing factors of the faults occurring on the molded part:*

   With the help of a cause-and-effect diagram (Figure 7.22), all possible influencing factors from Table 7.4 and Table 7.5 must now be sketched graphically. The advantage of this is that all the influencing factors you have determined that may have a certain effect on your molded part defect are broken down into their primary and secondary causes. This provides you with a clear, overall view for the upcoming inspections of the injection molding process and the injection mold

3. *Checking and evaluating all influencing factors of the process parameters and the injection mold:*

   The influencing factors from the injection molding process and injection mold, which are marked with a box (□), must be examined in detail. In addition to the already possible influencing factors of the machine setting parameters, the aim here is to identify the causes or sources of faults that occur via the process parameters and the injection mold

Table 7.4 Targeted Analysis of Influencing Factors from the Injection Molding Process

Fault on the molded part:	Machine parameters (Input)														Process parameters (Output)						
	Plasticizing phase					Injection phase			Switching to holding pressure	Holding pressure phase		Cooling phase	Mold		Plasticizing phase			Holding pressure phase		Cooling phas	Cavity pressure
	Temperature of the cylinder	Plasticizing path	Screw speed	Screw speed	Screw end compression	Injection pressure	Injection speed	Injection speed (stepped profile)	Switchover point	Holding pressure	Holding pressure time	Rest cooling time	Clamping force	Temperature of the mold	Dwell time of the melt in the cylinder	Temperature of the melt	Homogeneous melt (visual assessment)	Sealing time	Rest mass cushion	Demolding temperature	Evaluation of the curve progression
Optical faults:																					
Moisture streaks																	□				
Color streaks	X↑		X↓	X↑											□	□	□				
Burn streaks	X↓		X↓	X↓			X↓	X							□	□					
Glass fiber streaks	X↑						X↑							X↑		□	□				
Sink marks	X↑						X			X↑	□			X↑				□	□		
Gloss, gloss differences	X↑						X↑		□	X				X↑							

Table 7.4 Targeted Analysis of Influencing Factors from the Injection Molding Process (*continued*)

	Machine parameters (Input)														Process parameters (Output)						
	Plasticizing phase					Injection phase			Switching to holding pressure	Holding pressure phase		Cooling phase	Mold		Plasticizing phase			Holding pressure phase		Cooling phas	Cavity pressure
Fault on the molded part:	Temperature of the cylinder	Plasticizing path	Screw speed	Screw speed	Screw end compression	Injection pressure	Injection speed	Injection speed (stepped profile)	Switchover point	Holding pressure	Holding pressure time	Rest cooling time	Clamping force	Temperature of the mold	Dwell time of the melt in the cylinder	Temperature of the melt	Homogeneous melt (visual assessment)	Sealing time	Rest mass cushion	Demolding temperature	Evaluation of the curve progression
Optical faults:																					
Unmelted material in the molded part	X↑		X↓	X↑											□	□	□				
Weld lines, flow seams	X↑						X↑			X↑				X↑							
Free jet formation	X↑						X↓	X													
Diesel effect (burner)							X↓	X					X↓								
Air hooks/ air streaks				X↑	X↓		X↓	X													

Table 7.4 Targeted Analysis of Influencing Factors from the Injection Molding Process (*continued*)

	Machine parameters (Input)														Process parameters (Output)						
	Plasticizing phase					Injection phase			Switching to holding pressure	Holding pressure phase		Cooling phase	Mold		Plasticizing phase			Holding pressure phase		Cooling phas	Cavity pressure
Fault on the molded part:	Temperature of the cylinder	Plasticizing path	Screw speed	Screw speed	Screw end compression	Injection pressure	Injection speed	Injection speed (stepped profile)	Switchover point	Holding pressure	Holding pressure time	Rest cooling time	Clamping force	Temperature of the mold	Dwell time of the melt in the cylinder	Temperature of the melt	Homogeneous melt (visual assessment)	Sealing time	Rest mass cushion	Demolding temperature	Evaluation of the curve progression
Optical faults:																					
Visible ejector marks									□	X↓		X↑		X↓						□	□
Coating formation in the mold	X↓		X↓	X↓																	
Disc effect	X↑						X↑		□					X↑							□
Dark spots	X↑		X↓	X↓											□	□	□				
Dull spots in the sprue area	X↑						X↓	X													

Table 7.4 Targeted Analysis of Influencing Factors from the Injection Molding Process (*continued*)

	Machine parameters (Input)														Process parameters (Output)						
	Plasticizing phase					Injection phase			Switching to holding pressure	Holding pressure phase		Cooling phase	Mold		Plasticizing phase			Holding pressure phase		Cooling phas	Cavity pressure
Fault on the molded part:	Temperature of the cylinder	Plasticizing path	Screw speed	Screw speed	Screw end compression	Injection pressure	Injection speed	Injection speed (stepped profile)	Switchover point	Holding pressure	Holding pressure time	Rest cooling time	Clamping force	Temperature of the mold	Dwell time of the melt in the cylinder	Temperature of the melt	Homogeneous melt (visual assessment)	Sealing time	Rest mass cushion	Demolding temperature	Evaluation of the curve progression
Optical faults:																					
Flaking of the surface layer	X↑						X↓	X						X↑			□				
Cold plugs	X↑			X	X																
Incompletely filled parts	X↑					□	X↑		□					X↑							□
Molded part overmolded (flash formation, floating skin)	X↓							X	□	X↓			X↑	X↓				□			□

Table 7.4 Targeted Analysis of Influencing Factors from the Injection Molding Process (*continued*)

Fault on the molded part:	Machine parameters (Input)														Process parameters (Output)						
	Plasticizing phase					Injection phase			Switching to holding pressure	Holding pressure phase		Cooling phase	Mold		Plasticizing phase			Holding pressure phase		Cooling phas	Cavity pressure
	Temperature of the cylinder	Plasticizing path	Screw speed	Screw speed	Screw end compression	Injection pressure	Injection speed	Injection speed (stepped profile)	Switchover point	Holding pressure	Holding pressure time	Rest cooling time	Clamping force	Temperature of the mold	Dwell time of the melt in the cylinder	Temperature of the melt	Homogeneous melt (visual assessment)	Sealing time	Rest mass cushion	Demolding temperature	Evaluation of the curve progression
Optical faults:																					
Deformation during demoulding							X↓	X	□	X		X↑								□	□
Distortion of the molded part	X						X		□	X	□			X			□	□		□	□
Stress cracks, white crack on the moulded part	X						X	X	□	X↓		X		X							□

Table 7.4 Targeted Analysis of Influencing Factors from the Injection Molding Process (*continued*)

Fault on the molded part:	Machine parameters (Input)														Process parameters (Output)						
	Plasticizing phase					Injection phase			Switching to holding pressure	Holding pressure phase		Cooling phase	Mold		Plasticizing phase			Holding pressure phase		Cooling phas	Cavity pressure
	Temperature of the cylinder	Plasticizing path	Screw speed	Screw speed	Screw end compression	Injection pressure	Injection speed	Injection speed (stepped profile)	Switchover point	Holding pressure	Holding pressure time	Rest cooling time	Clamping force	Temperature of the mold	Dwell time of the melt in the cylinder	Temperature of the melt	Homogeneous melt (visual assessment)	Sealing time	Rest mass cushion	Demolding temperature	Evaluation of the curve progression
Optical faults:																					
Air inclusions, bubble formation	X↓		X↓	X↑											☐	☐	☐				
Blowholes/ vacuoles	X						X		☐	X↑	☐			X				☐	☐		☐
Thermal faults:																					
Insufficient tempera-ture control																X				☐	

Table 7.4 Targeted Analysis of Influencing Factors from the Injection Molding Process (*continued*)

Fault on the molded part:	Machine parameters (Input)														Process parameters (Output)						
	Plasticizing phase					Injection phase			Switching to holding pressure	Holding pressure phase		Cooling phase	Mold		Plasticizing phase			Holding pressure phase		Cooling phas	Cavity pressure
	Temperature of the cylinder	Plasticizing path	Screw speed	Screw speed	Screw end compression	Injection pressure	Injection speed	Injection speed (stepped profile)	Switchover point	Holding pressure	Holding pressure time	Rest cooling time	Clamping force	Temperature of the mold	Dwell time of the melt in the cylinder	Temperature of the melt	Homogeneous melt (visual assessment)	Sealing time	Rest mass cushion	Demolding temperature	Evaluation of the curve progression
Optical faults:																					
Hot spots on the molded part												X↑								□	
Heat pockets on the molded part												X↑								□	
Demolding temperature too high												X↑								□	

Table 7.5 Targeted Analysis of Influencing Factors from the Injection Mold

Fault on the molded part:	Injection Mold												
	Demolding				Thermal				Design/construction				
	Sharp edges or transitions	Unpolished ribs, breakthroughs	Undercut	Conicity (Entformungsschrägen)	Temperature distribution	Mold wall temperature	Hot runner system	Flow rate of temperature contol medium	Mold venting	Stability (supports, core supports, rigidity etc.)	Gating (cross-sectional area too small, high friction etc.)	Mold breathing	Touching parts
Optical faults:													
Moisture streaks													
Color streaks													
Burn streaks	☐						☐				☐		
Glass fiber streaks					☐	☐	☐						
Sink marks					☐	☐					☐		
Gloss, differences in gloss	☐	☐			☐	☐	☐		☐		☐		
Unmelted material in the molded part													
Weld line, flow line					☐	☐			☐				
Jetting											☐		
Diesel effect (burns)									☐		☐		☐
Air hooks/air streaks	☐								☐				
Visible ejector marks		☐	☐	☐									
Plate-out									☐				
Vinyl record effect											☐		
Dark spots							☐						
Dull spots in the gate area											☐		
Flaking of the surface layer													
Cold slug											☐		

Table 7.5 Targeted Analysis of Influencing Factors from the Injection Mold (*continued*)

Fault on the molded part:	Injection Mold												
	Demolding				Thermal				Design/construction				
	Sharp edges or transitions	Unpolished ribs, breakthroughs	Undercut	Conicity (Entformungsschrägen)	Temperature distribution	Mold wall temperature	Hot runner system	Flow rate of temperature contol medium	Mold venting	Stability (supports, core supports, rigidity etc.)	Gating (cross-sectional area too small, high friction etc.)	Mold breathing	Touching parts
Optical faults:													
Incompletely filled parts					☐	☐	☐		☐		☐		
Molded part over-molded (flash formation)										☐		☐	☐
Deformation during demolding	☐	☐	☐	☐						☐		☐	
Warpage of the molded part					☐	☐	☐	☐			☐		
Stress cracks, white cracks on the molded part	☐	☐	☐	☐	☐	☐				☐			
Air inclusions, bubble formation									☐				
Blowholes/ vacuoles											☐		
Thermal faults:													
Inadequate temperature control					☐	☐		☐					
Hot spots on the molded part					☐	☐							
Heat pockets on the molded part					☐	☐	☐	☐					
Demolding temperature too high					☐	☐	☐	☐					

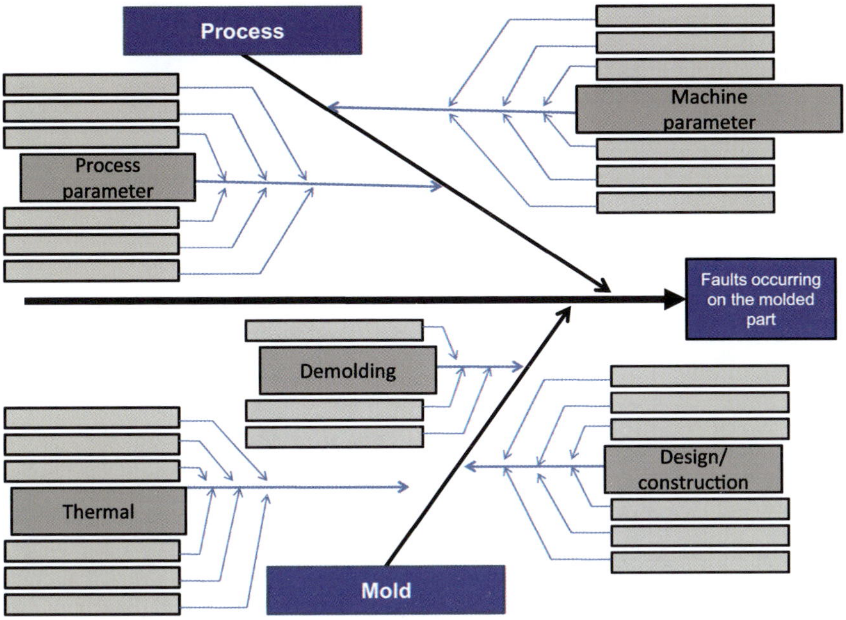

Figure 7.22 Summarizing in a cause-and-effect diagram

In order to successfully achieve the goal, the options described in detail, such as thermography, cavity pressure for detecting causes of the fault or sources of interference in the injection molding process and injection mold, must always be employed in the course of a mold sampling analysis, as follows:

General/optical inspections:

- Machine setting parameters (target/actual comparison)
- Evaluation of the general process parameters
- Residual melt cushion
- Dwell time of the melt in the plasticizing barrel
- Checking the dimensional accuracy
- Visual inspection of the injection mold (demolding, touching)
- Flow rate of the temperature control medium (venting, gating, etc.)

Thermal inspections (thermography):

- Demolding temperature (actual/target comparison)
- Melt temperature (actual/target comparison)
- Temperature distribution in the molded part and injection mold
- Temperature difference (ΔT) between the flow and return temperature of the temperature control unit

- Mold wall temperature
- Hot spots on the molded part (inadequate temperature control in the mold)
- Mold temperature control connected correctly and effectively

Check cavity pressure curve:

- Correct switchover point to holding pressure
- Optimal holding pressure level and holding pressure time
- Correct injection speed
- Do not demold under residual pressure

Step 4: Evaluation of the mold sampling analysis:

In order to be able to determine the next steps in the sampling process, a few questions must first be answered:

- Were tooling samples produced?
- Could these tooling samples be demolded without any problems?
- Are the faults occurring on the molded part exclusively due to the injection molding process?
- Are there no scratches, damage or "seizure marks" in the mold (e. g. on all moving parts such as slides/splits, core pulls, ejector pins, etc.)?
- Are there no serious defects in the injection mold, e. g. demolding, layout, etc.?

If all these questions can be answered with YES, the faults that occur on the molded part can be eliminated or improved with the help of systematic optimization of the injection molding process (Chapter 8 "Optimizing the basic setting").

If the answer is NO, the "mold sampling specialist" must decide to what extent systematic optimization of the injection molding process makes sense despite possible defects in the mold.

Unfortunately, during sampling, attempts are often made to use the machine setting parameters to eliminate or optimize faults in the molded part caused by defects in the injection mold. The reason for this approach is usually that corrections or optimizations to the mold are time-consuming and associated with additional costs. Unfortunately, this approach is not expedient.

An optimal injection molding process should have a large process window, which makes it possible to compensate for all influencing factors with the help of the machine and process parameters. If a setting is used to compensate for possible defects in the mold, the result is subsequent process fluctuation and process uncertainty in series production.

REMEMBER:

If there are faults in the injection mold during your sampling that could negatively influence the optimization of your basic setting, these must first be rectified in the mold before the basic setting can be optimized.

For this purpose, all defects occurring on the injection mold during the mold sampling analysis are documented in the sampling report (Chapter 9 "Documentation of mold sampling"). The basic setting is therefore optimized only after the mold correction during the second mold sampling. Figure 7.23 shows a schematic diagram of the subsequent mold sampling procedure.

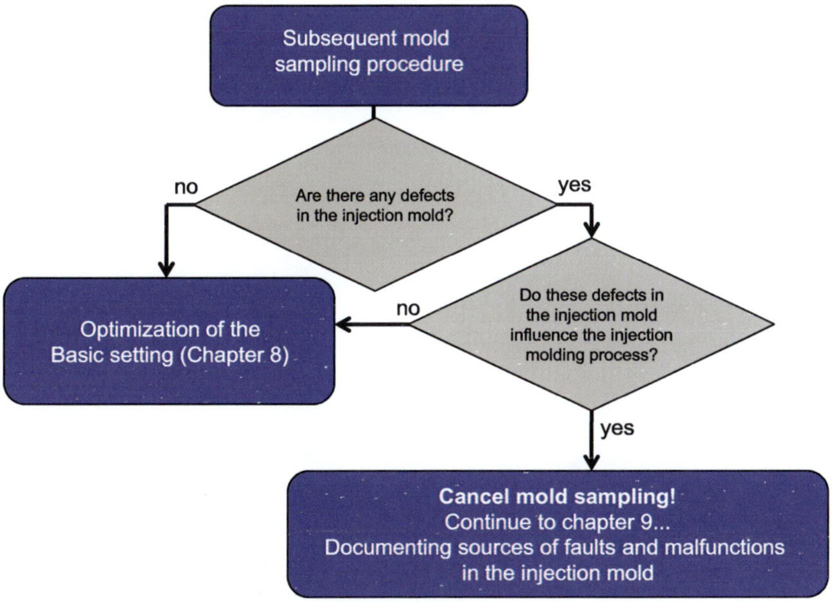

Figure 7.23 Defining the optimization strategy (schematic)

Milestone: "Mold sampling analysis of the basic setting"

- Block 6 of the sampling checklist has been processed in full
- The basic setting of the injection molding process has been analyzed
- All sources of faults and malfunctions in the injection molding process and injection mold have been localized, defined and examined in more detail
- The subsequent optimization strategy for the process and mold has been determined

References

[1] Johannaber, F., Michaeli, W.: Handbuch Spritzgießen, 2nd Edition, 2004, Carl Hanser Verlag, Munich
[2] Steinko, W.: Optimierung von Spritzgießprozessen, 2007, Carl Hanser Verlag, Munich

8 Optimization of the Basic Setting

Following the mold sampling analysis (Chapter 7 "Mold sampling analysis of the basic setting"), the next sampling step is to optimize the basic setting. Using this chapter, the "mold sampling specialist" systematically works through Block 8 ("Optimization of the basic setting") of the sampling checklist (Figure 8.1).

Optimization of the basic setting				
Optimization of the basic setting–Part 1:		**Yes**	**No**	**Block 8**
51	Optimization strategy defined?			
52	Molding tests carried out?			
53	Molding tests analyzed?			
Optimization of the basic setting–Part 2:		**Yes**	**No**	
54	Basic setting evaluated and optimized for productivity?			
55	Basic setting evaluated and optimized for energy efficiency?			
Optimization of the basic setting–Part 3:		**Yes**	**No**	
56	Process analysis of machine setting and process parameters carried out and evaluated?			
57	Run@Rate process of the optimized basic setting performed and evaluated?			

Figure 8.1 Mold sampling checklist: Block 8

This optimization is a very important phase of the sampling process, which is about using a structured approach to address and eliminate all the faults on the molded part and sources of interference in the injection molding process.

Optimizing an injection molding process is often difficult because it is non-linear, complex and multivariable.

This means that certain machine settings and process parameters can interact with each other. Major deficits in the approach as a result of a lack of process knowledge are responsible for improper optimization of the basic setting in practice. Frequent causes of faults are:

- Non-systematic changing of one or more machine setting parameters at the same time
- Missing correlation between machine setting or process parameters and part quality
- Absence of a setting or optimization strategy to eliminate faults
- Missing documentation and evaluation of injection molding tests
- No time for injection molding tests

In this chapter, I would therefore like to introduce you to a procedure for optimizing your current basic setting. I will introduce you to optimization methods that will help you achieve your goal step by step, quickly and effectively. The aim of any optimization of the basic setting must therefore be to eliminate all faults that occur in your molded part and to optimize the injection molding process, taking into account the high demands on cost-effectiveness and energy efficiency. Furthermore, a process capability analysis must be implemented at the end of the optimization phase with the aim of examining the spread of all quality-relevant machine setting and process parameters.

Based on the knowledge gained from this, you can define an optimum tolerance for the machine or process parameters to be monitored on the machine. For a better understanding, a practical example is given at the end of each optimization step.

PRACTICAL EXAMPLE:

Molded part: Housing cover

Plastic: PC/ABS

Faults on the molded part: Differences in gloss in the weld line area

Figure 8.2 Practical example: "Gloss differences in the weld line area" (image source: Gentner Kunststoffverarbeitung GmbH)

The procedure for optimizing the basic settings is clearly set out in Figure 8.3. All steps are described in more detail below.

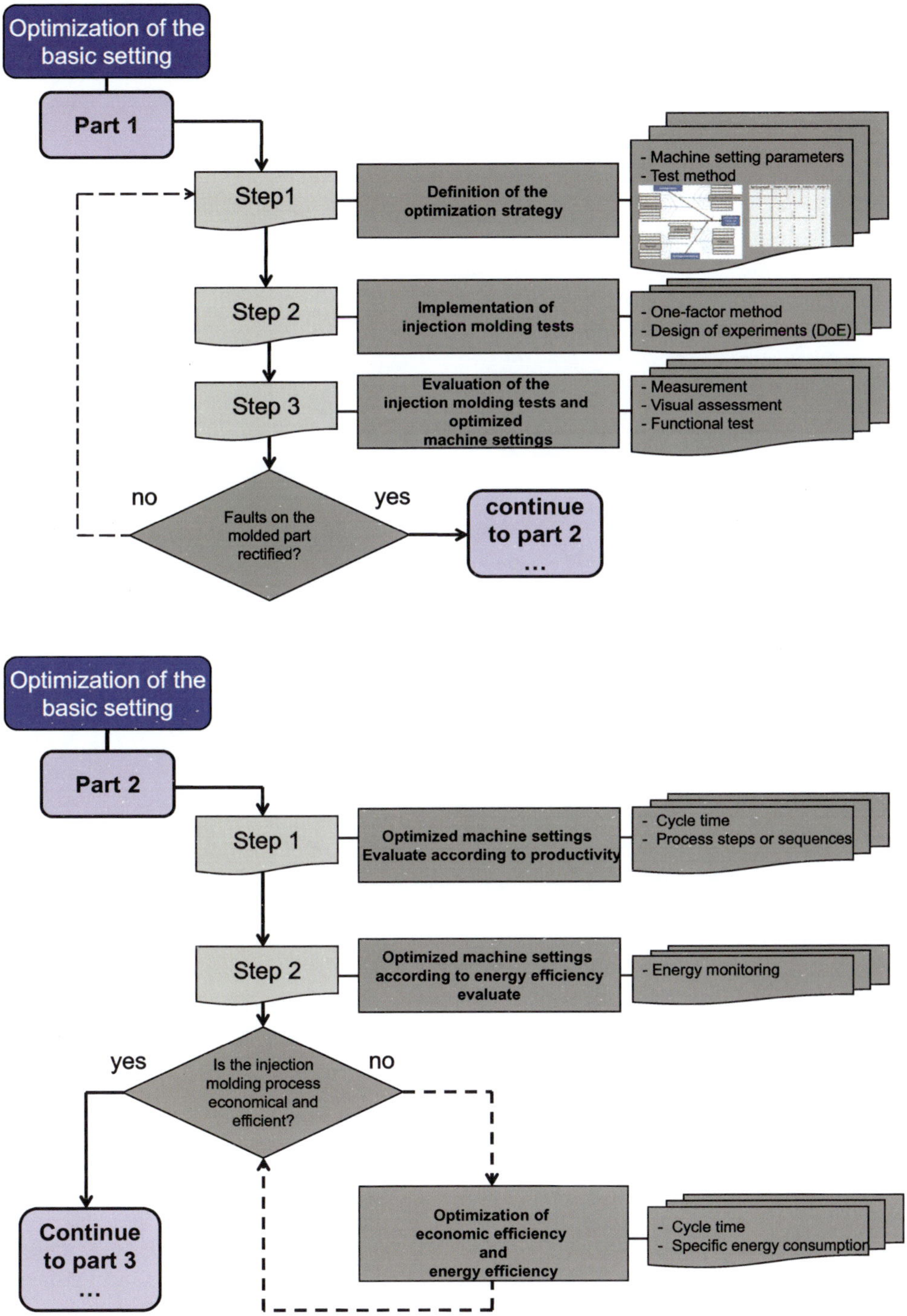

Figure 8.3 Procedure for optimizing the basic setting (schematic)

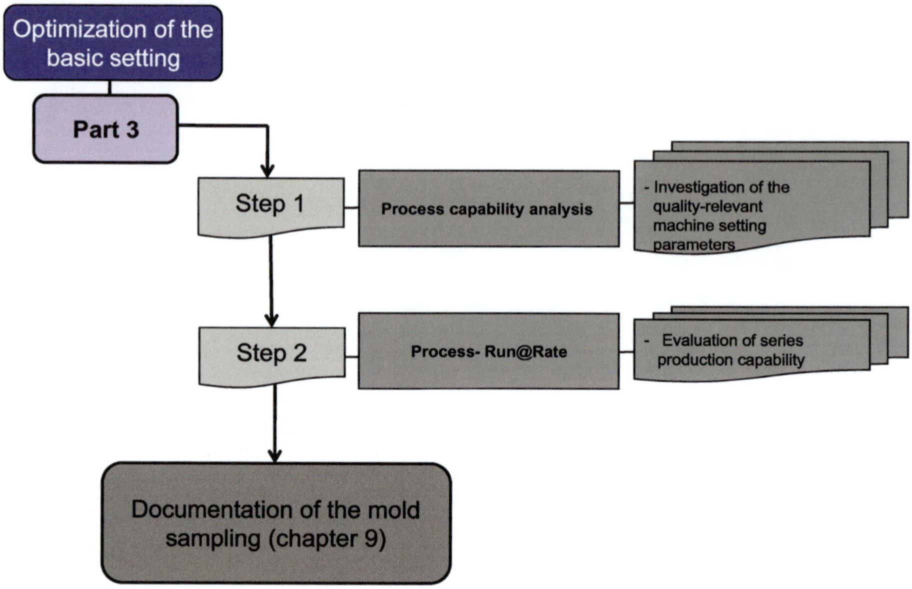

Figure 8.3 Procedure for optimizing the basic setting (schematic) (*continued*)

8.1 Optimizing the Basic Setting—Part 1

8.1.1 Step 1: Definition of the Optimization Strategy

Optimization of the basic setting begins with the definition of a strategy. This is very important so that you can implement an optimization of the injection molding tests from the outset using the correct test methodology.

In order to be able to define a strategy, you now need all the knowledge collected in the cause-and-effect diagram from the previous mold sampling analysis in Chapter 7 "Mold sampling analysis of the basic setting". These are:

- The faults occurring on the molded part
- Possible influencing factors (machine parameters: input) of the fault on the molded part (e.g. barrel temperature, holding pressure, injection speed, etc.)
- The knowledge to evaluate the necessary checking or optimization of possible process parameters (output) of the fault on the molded part (e.g. cavity pressure curve, dwell time of the melt, residual melt cushion, demolding temperature, etc.)
- The number of influencing factors (factor A, B, C)

Only then is it possible for you to make a decision based on the available knowledge as to which influencing factors of the machine setting parameters (input) and process parameters (output) should be included in the optimization or check. Once your influencing factors have been determined, nothing stands in the way of defining the test execution or optimization strategy.

PRACTICAL EXAMPLE:

In the case study, the following possible cause-and-effect diagram (Figure 8.4) emerges for the fault "gloss differences on the molded part in the weld line area".

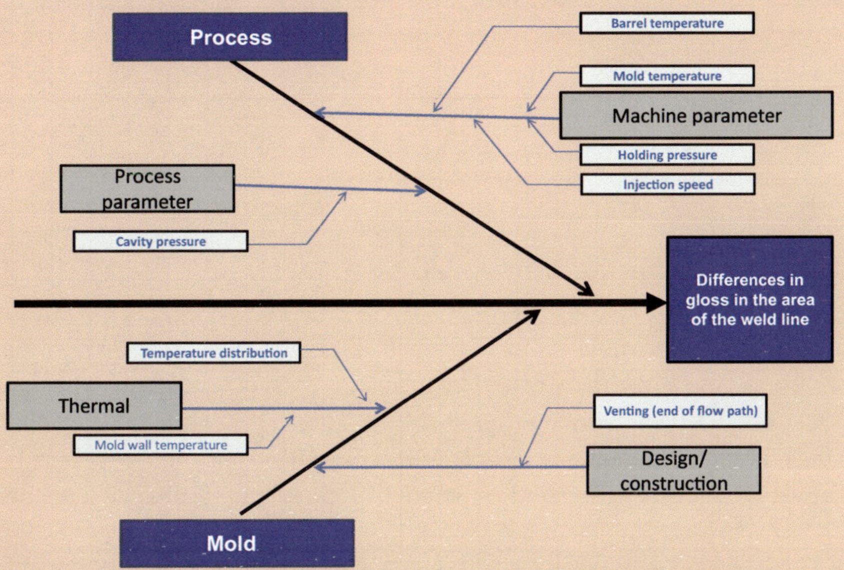

Figure 8.4 Cause-and-effect diagram for the example of "gloss differences in the weld line area"

Based on the knowledge gathered in the cause-and-effect diagram (Figure 8.4) the following important details can be extracted for the optimization of the basic setting:

- Faults occurring on the molded part: gloss differences in the weld line area
- Possible influencing factors (machine parameters (input)) of the fault occurring on the molded part: barrel temperature (°C), injection speed (mm/s), mold temperature (°C), holding pressure (bar)
- Necessary check or optimization of possible process parameters (output) of the fault on the molded part: cavity pressure curve
- Necessary check of possible thermal influencing factors of the fault on the molded part: mold temperature distribution, cavity temperature (°C)

Thermal inspection with a thermal imaging camera (thermography) ruled out a thermal cause in the injection mold. Furthermore, an inspection of the cavity pressure curve did not reveal any extreme deviations from the target curve.

In the practical example shown for "gloss differences in the weld line area", the following machine setting parameters are implemented to optimize the basic setting:

- Factor A: Barrel temperature (°C)
- Factor B: Injection speed (mm/s)
- Factor C: Mold temperature (°C)

Once the main influencing factors have been defined, the next step is to determine the optimization strategy. Here you must decide which test method you would like to use to optimize the basic setting. To this end, I would first like to present two practical methods in more detail:

- One factor at a time (OFAT)
- Design of experiments (DoE)

8.1.1.1 One-Factor-at-a-Time Method (OFAT)

Nowadays, the OFAT method is often used in practice to optimize the machine. With this method, the possible influencing factors of the faults occurring on the molded part are varied individually, one after the other, with all other factors being kept constant.

As the number of influencing factors increases, so does the number of failed attempts, as more precise effects, such as interactions between the influencing factors, cannot be accurately recognized. Consequently, a maximum number of three factors should not be exceeded when this method is employed.

This OFAT method is very easy to apply and is therefore frequently employed in practice. Table 8.1 shows the advantages and disadvantages of this method.

Table 8.1 Advantages and disadvantages of the OFAT method

Advantages	Disadvantages
Easy implementation of the method	Optimization of each factor separately; all other factors are kept constant.
	Interactions between the individual influencing factors are not recognized and taken into account.
	A maximum of three influencing factors can be used, as otherwise the method becomes uneconomical.
	High time outlay

8.1.1.2 Design of Experiments (DoE)

A proven method for injection molding is design of experiments (DoE). In the following "Background knowledge" information box, I would first like to answer the following questions:

- What is design of experiments (DoE)?
- Why perform design of experiments (DoE)?

BACKGROUND KNOWLEDGE:

What is design of experiments (DoE)?

Design of experiments or DoE refers to a number of different methods, such as the systematic planning, implementation and evaluation of tests. In injection molding, DoE is the best procedure for systematically optimizing machine settings. This method allows you to consider several influencing factors simultaneously, including the interactions between them, and to implement an optimization with the help of the test results.

Why perform design of experiments (DoE)?

In order to achieve a real improvement in quality, it is necessary to determine and optimize process-optimal settings for the influencing variables relevant to quality. Design of experiments is a very effective tool for obtaining the maximum amount of meaningful information with the smallest possible number of tests.

2^x factorial design of experiments (DoE)

A method derived from design of experiments and based on factorial design of experiments can be used to implement purely experimental DoE for optimizing the basic setting. With this method, you can obtain a maximum amount of meaningful information from your injection molding process from a small number of injection molding tests with a view to finding the optimum machine setting parameters for optimizing your injection molding process.

I will now describe this simplified method ("2^x factorial design") of factorial design of experiments in more detail. The big difference between 2^x factorial designs and the OFAT method is that the optimization can take place with simultaneous variation of all determined influencing factors of the faults occurring on your molded part with their additional effects, such as interactions. With the help of 2^x factorial designs, injection molding tests on the machine can be evaluated without the need for complex mathematical processes.

Depending on the faults occurring in your molded part, the tests can be evaluated by measurement (e. g. dimensional problems, warpage), visual inspection (e. g. optical faults) or a functional test (e. g. shoring) of the molded part. This enables you to show each injection molding test in direct correlation with a test result and to evaluate it effectively. As a result, it becomes clear which injection molding test leads to an improvement or a deterioration, whereupon a decision can be made on how to proceed. The advantages of the simplified 2^x factorial design are as follows:

- Structured and efficient implementation of injection molding tests on the machine
- Less testing effort and therefore less time required
- Determination of the main factors influencing the faults occurring on the molded part
- The correlation between machine setting and process parameters and the molded part quality becomes apparent
- Determination of the process variance of the machine setting or process parameters (target/actual comparison)
- No mathematical processes are necessary
- It is systematic: the factors are changed simultaneously

Set-up of a 2^x factorial design

First of all, depending on the number of possible influencing factors of your faults on the molded part from the mold sampling analysis (Chapter 7 "Mold sampling analysis of the basic setting"), you must decide which 2^x factorial design is best suited for your test execution. The test effort is thus known from the outset and can be selected individually. For this purpose, I have compiled an overview of tried and tested factorial designs (Table 8.2). Here the factors A, B, C, D stand for the respective influencing factor (machine setting parameters) of the fault occurring on your molded part. These individual factors are increased (+) or decreased (–) depending on injection molding test 1, 2, 3, ... etc. with the same setting values. Table 8.1 shows possible constellations of a 2^x factorial design for optimizing the basic setting.

Table 8.2 Overview of Possible 2^x Factorial Designs by Number of Factors

Injection molding test	Factor A	Factor B	Factor C	Factor D
1	–	–	–	–
2	+	–	–	–
3	–	+	–	–
4	+	+	–	–
5	–	–	+	–
6	+	–	+	–
7	–	+	+	–
8	+	+	+	–
9	–	–	–	+
10	+	–	–	+
11	–	+	–	+
12	+	+	–	+
13	–	–	+	+
14	+	–	+	+
15	–	+	+	+
16	+	+	+	+

Table 8.2 shows the following factorial designs:

- 2^1 factorial design:
  1 influencing factor (A) with 2 setting values = 2 injection molding tests
- 2^2 factorial design:
  2 influencing factors (A, B) with 2 setting values = 4 injection molding tests
- 2^3 factorial design:
  3 influencing factors (A, B, C) with 2 setting values = 8 injection molding tests
- 2^4 factorial design:
  4 influencing factors (A, B, C, D) with 2 setting values = 16 injection molding tests

8.1.2 Step 2: Performing Injection Molding Tests

The next step is to perform injection molding tests on the machine using the selected optimization strategy. Important points must be observed when implementing these tests:

- Minimum and maximum number of molded parts per molding test
- Run-in time between the individual molding tests
- Labeling of the molded parts
- Temperature differences during molding tests
- Correct packaging and storage of the molding tests

Minimum and maximum number of molded parts per test:

Experience has shown that 10 to 15 molded parts per injection molding test are sufficient for a meaningful evaluation. I advise you not to produce fewer parts per test, as too few molded parts will prevent you from being able to see any variation in the results between the individual molded parts of a test. However, more molded parts per injection molding test are uneconomical, as the production of the molded parts is very time-consuming.

Run-in time between the individual injection molding tests:

Each newly set injection molding test on the machine requires a certain running-in period and should last at least 10–15 injection cycles before you remove the molded parts to be evaluated later. The reason is that this is the only way to ensure that each effect on the molded part can be correctly assigned to each injection molding test.

Labeling of the molded parts:

As the injection molding tests are being performed, the respective molded parts from each test must be labeled. This can be done using labels or in writing, for example. The labeling should look like this:

Fault occurring on the molded part:	XXXXXXXX
Injection molding test/No.:	XXXXXXXXX
Molded part/No. per test:	XXXXXXX
Date of manufacture:	XXXXXXXX

Temperature differences during injection molding tests:

As is so often the case in injection molding, temperature tests are frequently included, such as melt and mold temperature. To avoid unnecessary waiting time until the mold has cooled down from 60 °C to 20 °C for the next injection molding test, I recommend that you first run all the tests at the colder temperatures and then all the tests at the warmer temperatures.

Correct packaging and storage of molded parts:

If there are faults in the molded part due to warpage or the exceeding of tolerances in respect of dimensional accuracy, these test parts must be measured later for evaluation. It is therefore particularly important that the molded parts be packaged and stored correctly. If these parts are squashed, compressed or stored on top of each other, this can influence the measured values.

Care should therefore always be taken to ensure that the molded parts are already packed in the transport boxes, cardboard boxes or plastic trays which will be used in later series production.

Now that the basic points for implementing the injection molding tests have been clarified, I would like to present the implementation in detail using the two test methods already presented, the OFAT method and design of experiments (DoE). The advantages of this are that routine procedures do not always have to be rethought and the risk of forgetting something important is minimized.

8.1.2.1 Performing Molding Tests by the OFAT Method

As already mentioned, the one-factor-at-a-time method is the simplest way to optimize an injection molding process. However, this method reaches its limits very quickly when it comes to more complex fault patterns on the molded part. An OFAT method template (Figure 8.5) has been created for the "mold sampling specialist" to ensure that the method is implemented correctly.

I would like to explain in more detail how injection molding tests are implemented using the OFAT method with the help of our case study in the following practical example.

PRACTICAL EXAMPLE:

A fault occurring on the molded part, in our example "gloss differences in the weld line area", depends on several influencing factors (e. g. barrel temperature, injection speed, mold temperature). To determine the main influencing factor, proceed as follows: When varying the first influencing factor, e. g. barrel temperature (↑), keep the other influencing factors constant. If an optimum barrel temperature is found, this is retained and the second possible influencing factor is varied in the same way, and so on. Figure 8.6 shows the OFAT method.

One-factor-at-a-time method					
Article description:					
Mold number:					
Machine operator:					
Date:					
Fault occurring on the molded part:					
Determined influencing factors and output variables:	Factor A:		Actual value:		
	Factor B:		Actual value:		
	Factor C:		Actual value:		

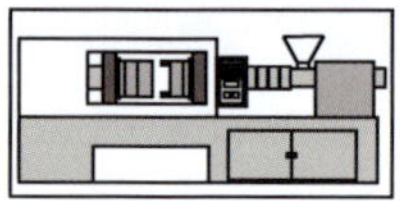

Variation of factor A (factor B and C constant)			
Factors:	A	B	C
Experiment 1			
Experiment 2			
Experiment 3			
Experiment 4			
Experiment 5			
Experiment 6			
Experiment 7			
Experiment 8			
Experiment 9			
Experiment 10			
best result (factor A):			

Variation of factor B (factor A and C constant)			
Factors:	A	B	C
Experiment 1			
Experiment 2			
Experiment 3			
Experiment 4			
Experiment 5			
Experiment 6			
Experiment 7			
Experiment 8			
Experiment 9			
Experiment 10			
best result (factor A and B):			

variation of factor B (factor A and C constant)			
Factors:	A	B	C
Experiment 1			
Experiment 2			
Experiment 3			
Experiment 4			
Experiment 5			
Experiment 6			
Experiment 7			
Experiment 8			
Experiment 9			
Experiment 10			
best result (factor A, B and C):			

Figure 8.5 The "OFAT method" template

One-factor-at-a-time method					
Article description:	Cover "Housing"				
Mold number:	11082006				
Machine operator:	XXXXXXX				
Datum:	XXXXXXX				
Fault occurring on the molded part:	Gloss differences in the weld line area				
Determined influencing factors and output variables:	Factor A:	Barrel temperature	Actual value:	°C	250
	Factor B:	Injection speed	Actual value:	mm/s	30
	Factor C:	Mold temperature	Actual value:	°C	50

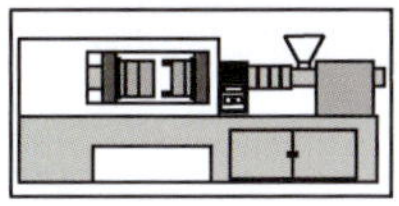

Variation of factor A (factor B and C constant)			
Factors:	A	B	C
Experiment 1	240	30	50
Experiment 2	265		
Experiment 3	270		
Experiment 4	275		
Experiment 5	280		
Experiment 6			
Experiment 7			
Experiment 8			
Experiment 9			
Experiment 10			
best result (factor A):	280		

Variation of factor B (factor A and C constant)			
Factors:	A	B	C
Experiment 1	280	35	50
Experiment 2		40	
Experiment 3		45	
Experiment 4		50	
Experiment 5		60	
Experiment 6			
Experiment 7			
Experiment 8			
Experiment 9			
Experiment 10			
best result (factor A and B):	280	60	

Variation of factor B (factor A and C constant)			
Factors:	A	B	C
Experiment 1	280	60	45
Experiment 2			55
Experiment 3			65
Experiment 4			75
Experiment 5			80
Experiment 6			
Experiment 7			
Experiment 8			
Experiment 9			
Experiment 10			
best result (factor A, B and C):	280	60	80

Figure 8.6 Practical example of the OFAT method

8.1.2.2 Implementing Injection Molding Tests with Design of Experiments (DoE)

Implementing injection molding tests with design of experiments (DoE) is the most efficient method of optimizing the basic setting, especially for complex faults on the molded part and in the injection process. Here, too, I would first like to introduce you to the design of experiments (DoE) template.

I would like to explain how injection molding tests are implemented with the factorial design (DoE) in more detail with the help of our case study in the following practical example.

PRACTICAL EXAMPLE:

Definition of the test method:

Based on the three influencing factors determined in the first step "Definition of the optimization strategy", the 2^3 factorial design with two setting values and a total of eight injection molding tests is now applied. The task is to find out experimentally which influencing factor has the greatest influence on the fault that occurs and which setting constellation of the influencing variables (factors A, B, C) can improve or eliminate the differences in gloss on the molded part.

Test template:

Use the template (Figure 8.8) and enter the general data ① at the top of the header. In the template, also make a note of the fault occurring in your molded part ②. Based on the mold sampling analysis, there are three factors in our example, so the box with the 2^3 factorial design ③ is marked in the template. Next, enter the determined factors (machine setting parameters) and note the actual values ④. Now comes the definition of the test values. Here, the "mold sampling specialist" must decide individually by how much to increase (+) or decrease (–) the value in the test phase ⑤. Once you have defined plus and minus values for factors A, B, C, these are now added or subtracted from the actual value by the defined test value, depending on the test sequence, and entered ⑥.

Factorial design of experiments		
Article description:		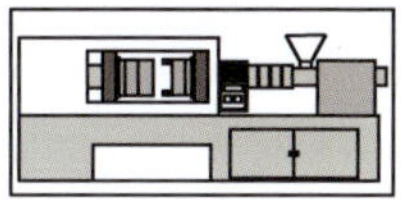
Mold number:		
Machine operator:		
Date:		
Fault occurring on the molded part:		
Performed experimental design:	☐ 2^1 Factorial design of experiments ☐ 2^2 Factorial design of experiments	☐ 2^3 Factorial design of experiments ☐ 2^4 Factorial design of experiments

Determined influencing factors and initial values:	Factor A:		Actual value:	
	Factor B:			
	Factor C:			
	Factor D:			

			Actual value:		(+) Increase	(-) Lower
Definition of test values:	Factor A:					
	Factor B:					
	Factor C:					
	Factor D:					

	Test number	Factor A		Factor B		Factor C		Factor D	
Test performance:	T1	(-)		(-)		(-)		(-)	
	T2	(+)		(-)		(-)		(-)	
	T3	(-)		(+)		(-)		(-)	
	T4	(+)		(+)		(-)		(-)	
	T5	(-)		(-)		(+)		(-)	
	T6	(+)		(-)		(+)		(-)	
	T7	(-)		(+)		(+)		(-)	
	T8	(+)		(+)		(+)		(-)	
	T9	(-)		(-)		(-)		(+)	
	T10	(+)		(-)		(-)		(+)	
	T11	(-)		(+)		(-)		(+)	
	T12	(+)		(+)		(-)		(+)	
	T13	(-)		(-)		(+)		(+)	
	T14	(+)		(-)		(+)		(+)	
	T15	(-)		(+)		(+)		(+)	
	T16	(+)		(+)		(+)		(+)	

Figure 8.7 The "Factorial design of experiments (DoE)" templated

Factoral design of experiments

Article description:	Cover "Housing"	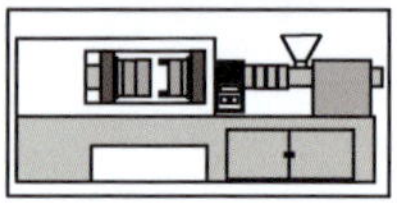
Mold number:	11082006 (1)	
Machine operator:	xxxxxx	
Date:	xxxxxx (2)	
Fault occurring on the molded part:	Gloss differences in the weld line area	
Performed experimental design:	☐ 2^1 Factorial design of experiments ☐ 2^2 Factorial design of experiments (3)	☒ 2^3 Factorial design of experiments ☐ 2^4 Factorial design of experiments

			Actual value:	
Determined influencing factors and initial values:	Factor A:	Barrel temperature (4)	250	°C
	Factor B:	Injection speed	30	mm/s
	Factor C:	Mold temperature	50	°C
	Factor D:			

			Actual value:		(+) Increase (5)	(−) Lower
Definition of test values:	Factor A:	Barrel temperature	250	°C	30	10
	Factor B:	Injection speed	30	mm/s	30	10
	Factor C:	Mold temperature	50	°C	30	10
	Factor D:					

	Test number	Factor A		Factor B		Factor C		Factor D	
Test performance:	T1	(−)	240	(−)	20	(−)	40	(−)	
	T2 (6)	(+)	280	(−)	20	(−)	40	(−)	
	T3	(−)	240	(+)	60	(−)	40	(−)	
	T4	(+)	280	(+)	60	(−)	40	(−)	
	T5	(−)	240	(−)	20	(+)	80	(−)	
	T6	(+)	280	(−)	20	(+)	80	(−)	
	T7	(−)	240	(+)	60	(+)	80	(−)	
	T8	(+)	280	(+)	60	(+)	80	(−)	
	T9	(−)		(−)		(−)		(+)	
	T10	(+)		(−)		(−)		(+)	
	T11	(−)		(+)		(−)		(+)	
	T12	(+)		(+)		(−)		(+)	
	T13	(−)		(−)		(+)		(+)	
	T14	(+)		(−)		(+)		(+)	
	T15	(−)		(+)		(+)		(+)	
	T16	(+)		(+)		(+)		(+)	

Figure 8.8 Practical example of 2^3 factorial design (DoE)

8.1.3 Step 3: Evaluation of the Injection Molding Tests

Once the injection molding tests have been successfully performed on the machine, the next step is to evaluate them. Here, the evaluation should not be carried out by the "mold sampling specialist" alone, but exclusively in a "four-eyes principle" together with the "quality assurance/metrology specialist". The four-eyes principle has the great advantage that the evaluation of the quality characteristics of the molded part is not conducted by one person.

The evaluation of the injection molding tests can look very different and depends on the fault on the molded part. Depending on the fault pattern on the molded part, the following three evaluation methods may be carried out separately or in combination with each other:

- **Measurement of the molded parts** by conventional methods, such as calipers, micrometers, height gauges, etc. or with a coordinate measuring machine or 3D scan.

> **NOTE:**
>
> Due to the shrinkage properties (post-shrinkage) of your molded parts, they should not be measured immediately after production.
>
> Sufficient resting time of the molded parts should be allowed in accordance with the raw material manufacturer's specifications or customer specifications. If no data is available, I recommend storing the molded parts in a standard climate (23 °C, 50% humidity) for at least 24 hours before measuring them and then measuring them in a standard climate.

- **Visual assessment** by comparing color and gloss level measurements, etc.
- **Functional test**, e. g. assembly tests

The first step here is to show the results from the individual injection molding tests in a transparent manner. This is the only way to evaluate the test results by direct comparison with each other and subsequently to assess the correlations between machine setting and process parameters and their effects on the molded part. In order to make your evaluation of the injection molding tests transparent, I recommend that you use a pre-prepared template.

8.1.3.1 Template for Evaluating the Injection Molding Tests

The template should be of a flexible design such that it can be used in accordance with the evaluation method and the number of injection molding tests carried out. I therefore recommend that you create an Excel template that is set up with a maximum of 16 trials (2^4 factorial design). This allows you to implement the evaluation directly in digital form using a laptop at the injection molding machine. The template for evaluating the injection molding tests should be set up as shown in Figure 8.9.

Evaluation of the injection molding tests

Article description:	
Mold number:	
Machine operator:	
Quality assurance/metrology specialist:	
Date:	
Fault occurring on the molded part:	

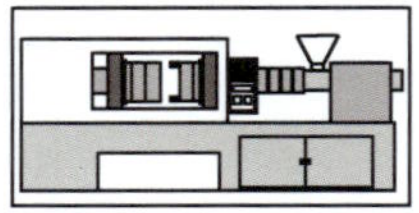

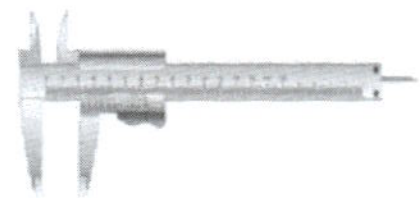

Factoren / Evaluation method	Injection molding test 1		Injection molding test 2		Injection molding test 3		....
	Factor A:	Value:	Factor A:	Value:	Factor A:	Value:	
	Factor B:	Value:	Factor B:	Value:	Factor B:	Value:	
	Factor C:	Value:	Factor C:	Value:	Factor C:	Value:	
	Factor D:	Value:	Factor D:	Value:	Factor D:	Value:	

Measurement	X-Direction	Nominal dimension	Tolerance	Part 1	Part 2	Part 3	Part 4	Part 5	Part 6	Average value	Part 1	Part 2	Part 3	Part 4	Part 5	Part 6	Average value	Part 1	Part 2	Part 3	Part 4	Part 5	Part 6	Average value	
	Y-Direction	Nominal dimension	Tolerance	Part 1	Part 2	Part 3	Part 4	Part 5	Part 6	Average value	Part 1	Part 2	Part 3	Part 4	Part 5	Part 6	Average value	Part 1	Part 2	Part 3	Part 4	Part 5	Part 6	Average value	
	Z-Direction	Nominal dimension	Tolerance	Part 1	Part 2	Part 3	Part 4	Part 5	Part 6	Average value	Part 1	Part 2	Part 3	Part 4	Part 5	Part 6	Average value	Part 1	Part 2	Part 3	Part 4	Part 5	Part 6	Average value	
Visual assessment	Note:			Part 1	Part 2	Part 3	Part 4	Part 5	Part 6	Medium result	Part 1	Part 2	Part 3	Part 4	Part 5	Part 6	Medium result	Part 1	Part 2	Part 3	Part 4	Part 5	Part 6	Medium result	
	Deterioration Neutral - No change Improvement																								
Function test	Note:			Part 1	Part 2	Part 3	Part 4	Part 5	Part 6	Medium result	Part 1	Part 2	Part 3	Part 4	Part 5	Part 6	Medium result	Part 1	Part 2	Part 3	Part 4	Part 5	Part 6	Medium result	
	Function not given Function given																								

Figure 8.9 The "Evaluation of injection molding tests" template

An evaluation template also offers you the following advantages during the sampling process:

- Optimal documentation of the injection molding tests carried out
- Recognition of interrelationships (correlations) between the injection molding test and the molded part
- Rapid, targeted evaluation
- Test procedure is easy for third parties to understand
- A "lessons learnt" process (systematic collection, evaluation and consolidation of empirical values for further trials)

With the help of the case study "gloss differences in the weld line area" on the "PC/ABS housing cover" molded part, I would like to show you how to work with such a template in more detail in the following practical example.

PRACTICAL EXAMPLE:

In our case study, we are dealing with a classic optical fault on the molded part, so the second step in the evaluation of the eight injection molding tests (2^3 experimental design) is a visual inspection. As this is a technical molded part, the focus is on dimensional accuracy as an important quality feature. For this reason, the individual molded parts of the tests are measured promptly. In this way, dimensional effects of the individual tests, such as changes or fluctuations, can be checked and included as a selection criterion in the evaluation.

Now take the "Evaluation of injection molding tests" template (Figure 8.10) and enter the general data ① at the top. The next step is to enter the factors and their value for the respective test (A, B, etc.) of your injection molding tests from your selected 2^3 factorial design (Figure 8.8) ②. As already known, our case study involves an optical fault on the molded part. Due to the high dimensional requirements for the molded part "PC/ABS housing cover", a prompt 3D measurement was carried out by the metrology department. Based on our example, select the evaluation method "Measurement" and "Optical assessment" ③. Using a measurement report from the quality assurance/metrology department, you must now assign the individual measurement results to your injection molding tests and document them under "Measurement" ④. You can now simultaneously evaluate a correlation to the dimensional accuracy of your molded parts for the "visual assessment" of the individual injection molding tests. The next step is to visually evaluate the molded parts. An evaluation in the example is carried out using a traffic light system. This means that the visual assessment is subject to three evaluation options, which the "quality assurance/metrology specialist" and the "mold sampling specialist" use to enter an evaluation of the respective injection molding test ⑤.

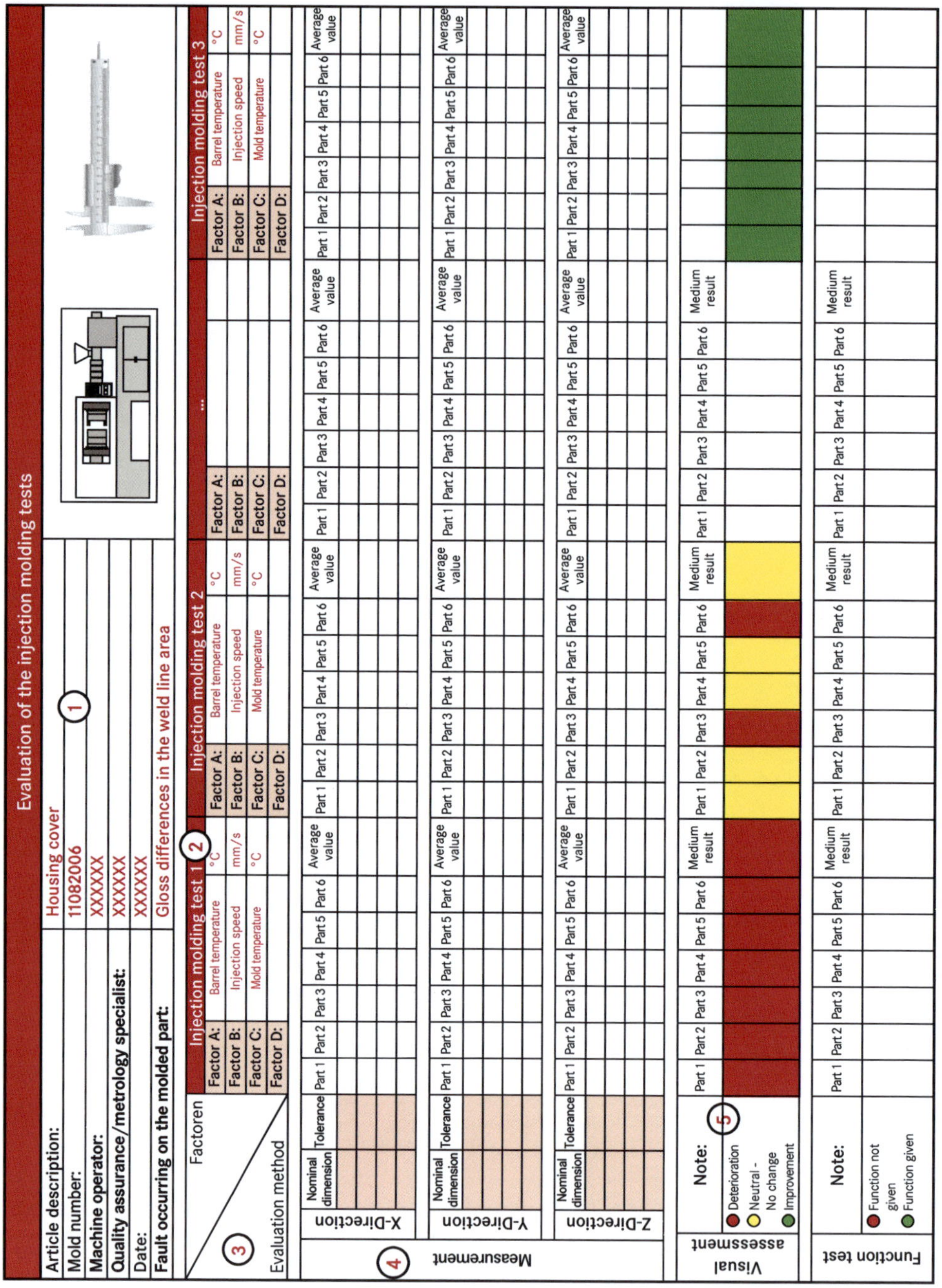

Figure 8.10 Practical example of an evaluation of injection molding tests

8.1.3.2 Definition of the Optimized Machine Settings

The evaluation of the injection molding tests is followed by the definition of the best injection molding test. This definition is determined with the help of a ranking of the individual evaluated injection molding tests. If the fault was corrected that occurred on the molded part, this machine setting is used as the optimized setting in the further course of the "Optimization of the basic setting".

If only a slight improvement in the fault on the molded part, e. g. visual appearance, dimensional accuracy and warpage, can be determined on the basis of the factorial design, you can continue to work with the actual values of the best injection molding test and start a new factorial design in order to gradually approach the optimum until the fault has been eliminated.

PRACTICAL EXAMPLE:

Summary of the practical example:

The fault "gloss difference in the weld line area" was corrected purely experimentally on the machine using a 2^3 factorial design. The fault pattern was eliminated by measurement and visual inspection. The evaluation of the injection molding tests clearly showed that injection molding test eight with a higher barrel temperature (280 °C), higher mold temperature (80 °C) and a faster injection speed of 60mm/s produced the best optical result. Furthermore, no negative correlation with the dimensional accuracy of the molded part could be determined by the injection molding test eight and thus the setting was adopted as the "optimized basic setting".

Figure 8.11 Optimized case study: "Gloss differences in the weld line area" (image source: Gentner Kunststoffverarbeitung GmbH)

8.2 Optimizing the Basic Setting—Part 2

Following the optimization of faults occurring on the molded part (Part 1), the second part now involves evaluating the optimized basic setting for economic efficiency and improving it if necessary.

In connection with economic efficiency, I would like to go into more detail on two key topics in the following sections:

- Productivity of the injection molding process
- Energy efficiency of the injection molding process

There is often a direct correlation between the productivity and the energy efficiency of the injection molding process (Figure 8.12). This means, for example, that a shorter residual cooling time has an influence on the cycle time and the specific energy consumption.

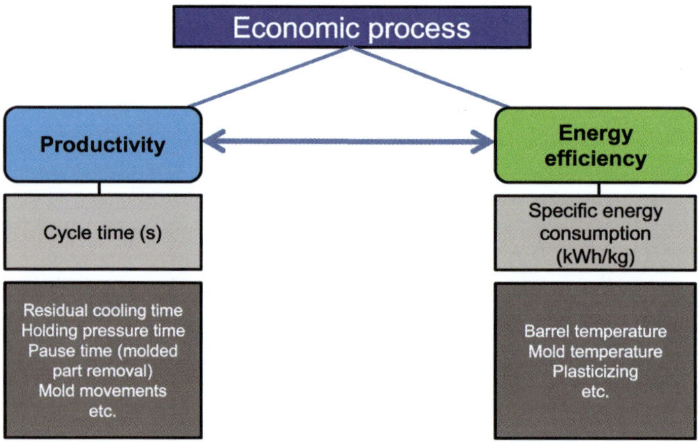

Figure 8.12 Productivity and energy efficiency of the injection molding process are often directly correlated

It is therefore particularly important to adjust the process so that there is a healthy balance in terms of productivity and energy efficiency while maintaining the quality of the molded parts.

> **NOTE:**
> A productive and energy-efficient optimization of the basic setting must not adversely affect the required quality characteristics of the molded part in any way.

8.2.1 Step 1: Evaluate and Optimize Optimized Basic Settings for Productivity

Productivity (or technical efficiency) is measured by the ratio of the quantitative yield (output) to the quantitative use of production factors (input). Consequently, the productivity of your injection molding process indicates how many molded parts can be produced in a given time.

It is therefore important to evaluate your optimized basic settings for optimum productivity and to improve them if necessary. The aim is to achieve an improved output/input ratio by increasing the number of molded parts (output) while strictly observing the quality characteristics of a molded part and at the same time reducing the cycle time (input).

In general, a possible increase in the productivity of each injection molding process should therefore be investigated. For example, a molded part can be produced in a cycle time of 50 seconds if the process is not optimized. That would be 72 molded parts in one hour. However, if just four seconds of cycle time were to be saved by optimizing productivity, that would be 78 parts in one hour. That is just six more molded parts per hour, but I would like to continue with this simplistic calculation for a moment.

As an example, we now have a series order for 10,000 housing covers. A non-optimized injection cycle of 50 seconds would have to produce for approx. 139 hours (≈ 5.78 days). However, if these four seconds could be saved in an optimization during the sampling phase, only 128.20 hours (≈ 5.34 days) would be needed to produce the 10,000 housing covers.

That's ten hours of time saved per series production alone, which you can use for further resource planning. You can see the potential that is made possible with even the smallest optimization of the productivity of your injection molding process. In the following section, I would therefore like to look at how you can take a systematic approach to an injection molding process and which influencing factors are the best levers for achieving an increase in productivity.

8.2.1.1 Optimization Options for Increasing Productivity

Every injection molding process has its optimization potential somewhere in the process sequence, but it is very important to ensure that the required quality level of the molded part is never adversely influenced or even changed when optimizing the cycle time.

REMEMBER:

A productive injection molding process produces a large number of parts in a short cycle time and in a consistent molded part quality.

Possibilities for increasing productivity in the injection molding process are shown in the following bar chart (Figure 8.13) using the individual process sequences over time.

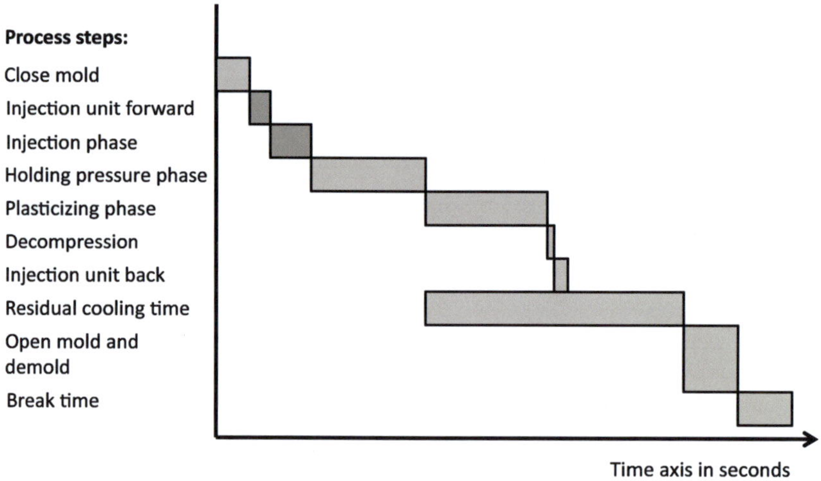

Figure 8.13 Bar chart of the injection molding cycle

Before an optimization strategy to increase the productivity of an injection molding process was implemented, the process steps shown in the bar chart (Figure 8.13) were divided into the following two categories:

- Process steps without direct influence on the molded part quality:
  - Close mold
  - Move the unit forward and backward
  - Open the mold and demold
  - Molded part removal
- Process steps with a direct influence on the molded part quality:
  - Injection phase
  - Holding pressure phase
  - Plasticizing phase
  - Decompression
  - Residual cooling

Process steps without direct influence on the molded part quality

The first step is to evaluate these process steps in more detail and, if possible, to improve the timing of these sequences in the injection molding process. The focus here is on mold movements such as strokes and speeds and the removal of molded parts (break), for example via handling. The "mold sampling specialist" should first check the following points of the individual process steps and optimize them if necessary:

- Is the set required opening stroke (mm) of the mold correctly selected?
- Is the speed profile (mm/s) of the mold movements adequate?
- Is the ejector stroke (mm) too long?
- Can the timing of the molded part removal (break) be improved?

Faster and shorter clamping unit strokes shorten the injection cycle. However, these can also be responsible for unsafe mold guidance that is not gentle on the mold if the process steps are too fast. A compromise should therefore be found between productivity and mold guidance that is gentle on the mold.

Process steps with a direct influence on the molded part quality

The holding pressure phase, the plasticizing phase and the residual cooling of the molded part clearly emerge from the bar chart (Figure 8.13) as the most time-consuming process steps. Shortening the holding pressure phase is definitely out of the question here, as the time required to seal the gate to the molded part is needed to compensate for the volume contraction. Shortening of the plasticizing phase should only be improved if the latter is set too slowly and predetermines the injection molding cycle after the required residual cooling time has elapsed. Therefore, the plasticizing phase should generally always be completed before the end of the residual cooling time.

The residual cooling of the molded part in the injection molding cycle takes up most of the time, so this process step must be correctly evaluated and optimized if necessary. A preliminary residual cooling time for the molded part has already been described in detail in Chapter 4 "Basic setting of the clamping unit". It is now necessary to check whether your provisionally set residual cooling time is sufficient to ensure the required demolding temperature of the molded part. This check is carried out with the thermal imaging camera (thermography) or, if no camera is available, with a thermometer. As the residual cooling time has a major qualitative influence on your molded part, the following steps must be observed when implementing the evaluation and optimization:

- **Step 1:**

  The temperature of a molded part is measured directly after demolding at the thickest point with the thermography camera or a thermometer

- **Step 2:**

  The measured demolding temperature of the molded part must now be compared with the required maximum demolding temperature (according to the raw material manufacturer). If the measured molded part temperature is below the recommended demolding temperature, this means, for example, that the preliminary residual cooling time was selected too long and must be optimized

- **Step 3:**

  Now reduce the residual cooling time in small steps and measure the molded parts for each reduction in residual cooling time until the raw material manufacturer's recommended demolding temperature is reached in all areas of the molded part

The optimization of the set residual cooling time is shown again below with the help of a simple practical example.

PRACTICAL EXAMPLE:

In Chapter 5 (Section 5.4.3.2 "Estimating using an approximation equation"), a residual cooling time of approx. 20 s was determined for the molded part housing cover using an approximation equation and set as the preliminary residual cooling time on the machine. In this case study, we assume a recommended demolding temperature of 105 °C. The first step is to investigate which demolding temperature is present as the initial temperature in the molded part. For this purpose, a thermal image of the molded part was captured immediately after demolding with a residual cooling time of 20 s (Figure 8.14).

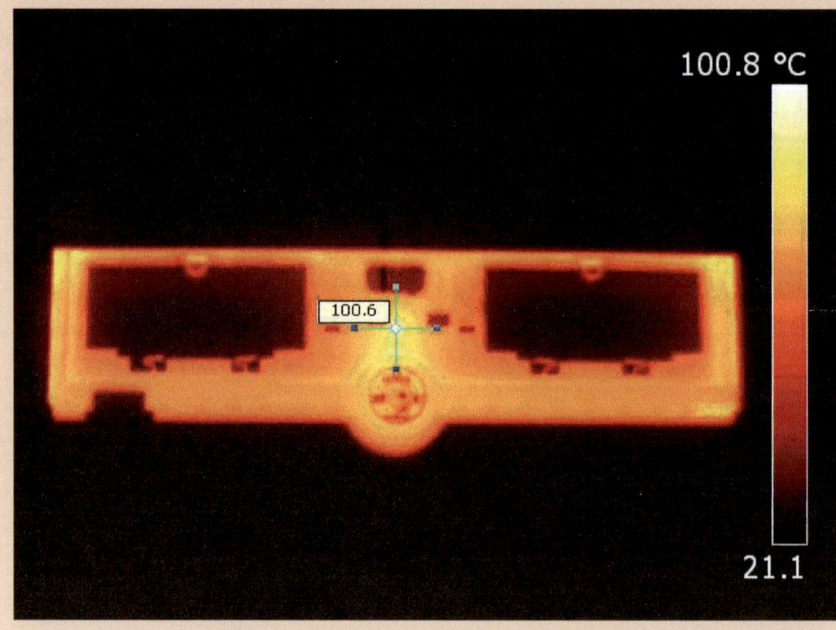

Figure 8.14 Thermal image with a residual cooling time of 20 s (starting situation)

In the practical example, the thermal imaging of the basic setting showed that the recommended demolding temperature was undercut by approx. 5 °C and therefore a residual cooling time optimization was required in order to achieve a reduction in cycle time and consequently an increase in the productivity of the injection molding process.

Based on this knowledge, the residual cooling time was corrected downward in small steps and the molded part was inspected in parallel with the thermography camera (Figure 8.15). As the required demolding temperature in the practical example is 105 °C, the residual cooling time was reduced from 20 s to 15 s, resulting in a cycle time reduction of 5 s in the injection process.

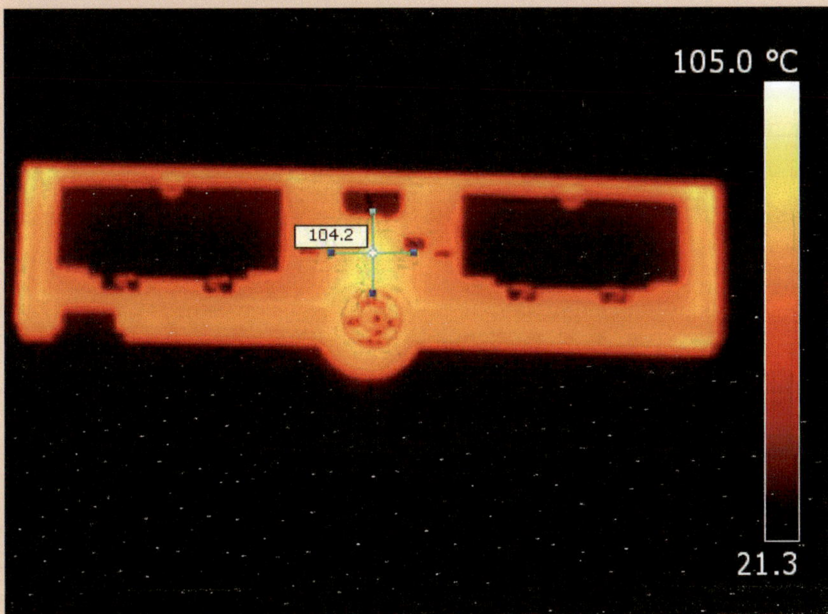

Figure 8.15 Thermographic image after shortening of the residual cooling time by five seconds

Based on this simple case study, you can now see that there is often great potential for increasing productivity by taking a systematic look at the injection molding process. Therefore, an evaluation of the productivity of the injection molding process should not be ignored in any mold sampling process.

8.2.2 Step 2: Evaluate and Optimize Optimized Basic Settings for Energy Efficiency

In addition to productivity, ever-increasing energy costs make it necessary to save energy. A basic setting should therefore also be optimized or evaluated in terms of energy efficiency.

Nowadays, the recording and documentation of production data for quality assurance of molded parts is a matter of course in companies. In practice, however, detailed recording and evaluation of the energy required in the injection molding process is often very rare or even non-existent. Without this recording and evaluation of the required energy demand of the injection molding machine, it is not possible to optimize the energy efficiency accordingly. In order to be able to make a targeted statement on the energy efficiency of the optimized basic setting, the energy consumption (kWh) of the injection molding process must first be measured with the help of “energy monitoring”.

BACKGROUND KNOWLEDGE:

What is energy monitoring?

Energy monitoring forms the basis for checking and optimizing the energy consumption of an injection molding machine. Energy monitoring is used to record electricity consumption and show it in digital or graphical form.

How does energy monitoring work?

Energy monitoring (Figure8.16) measures the current (A) and voltage (V) in all three phases as well as the phase shift. The measurement is carried out over a period of time X (h). This results in the apparent, active and reactive power in kW. As the energy consumption is the ratio of active power to measuring time, this gives the power consumption in kW/h.

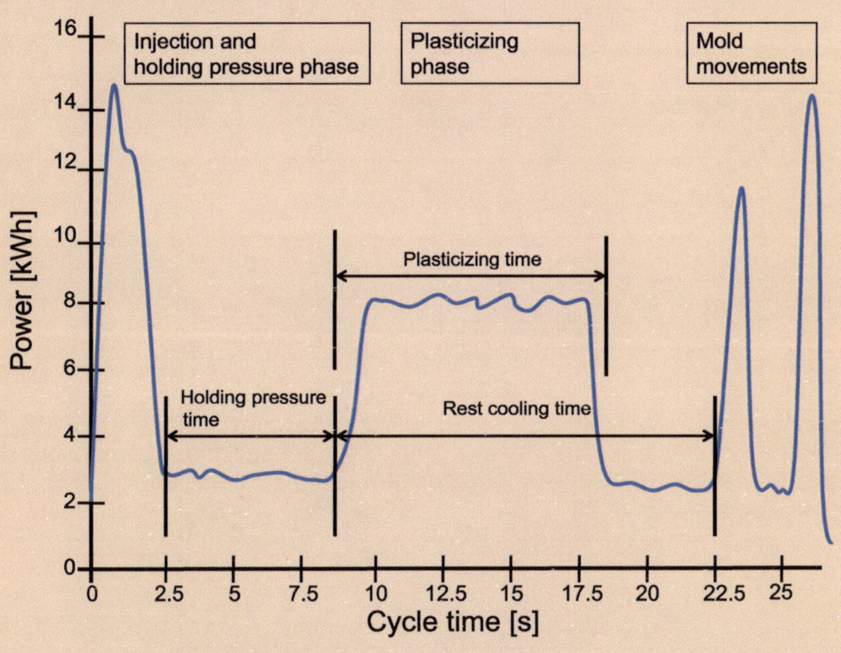

Figure 8.16 Example of energy monitoring of an injection molding process

What energy monitoring options are available?

The measured power consumption can nowadays be displayed as an optional additional function in the machine control system of modern injection molding machines. With the help of integrated energy monitoring in the control system, you can display the power consumption of the drives and heating, the current specific energy consumption and the individual consumptions in the injection molding cycle, as shown in (Figure 8.17).

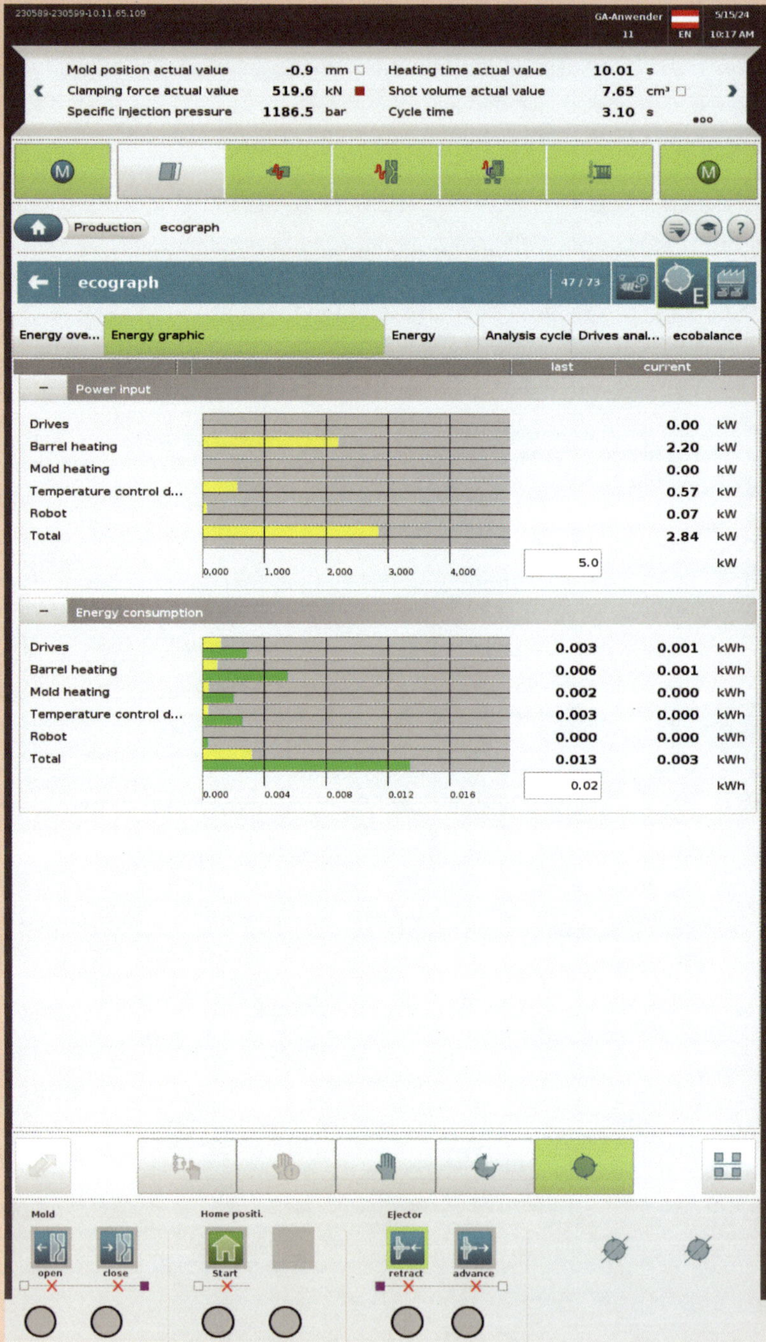

Figure 8.17 Energy monitoring in the machine controller (image source: Ecograph, Engel GmbH

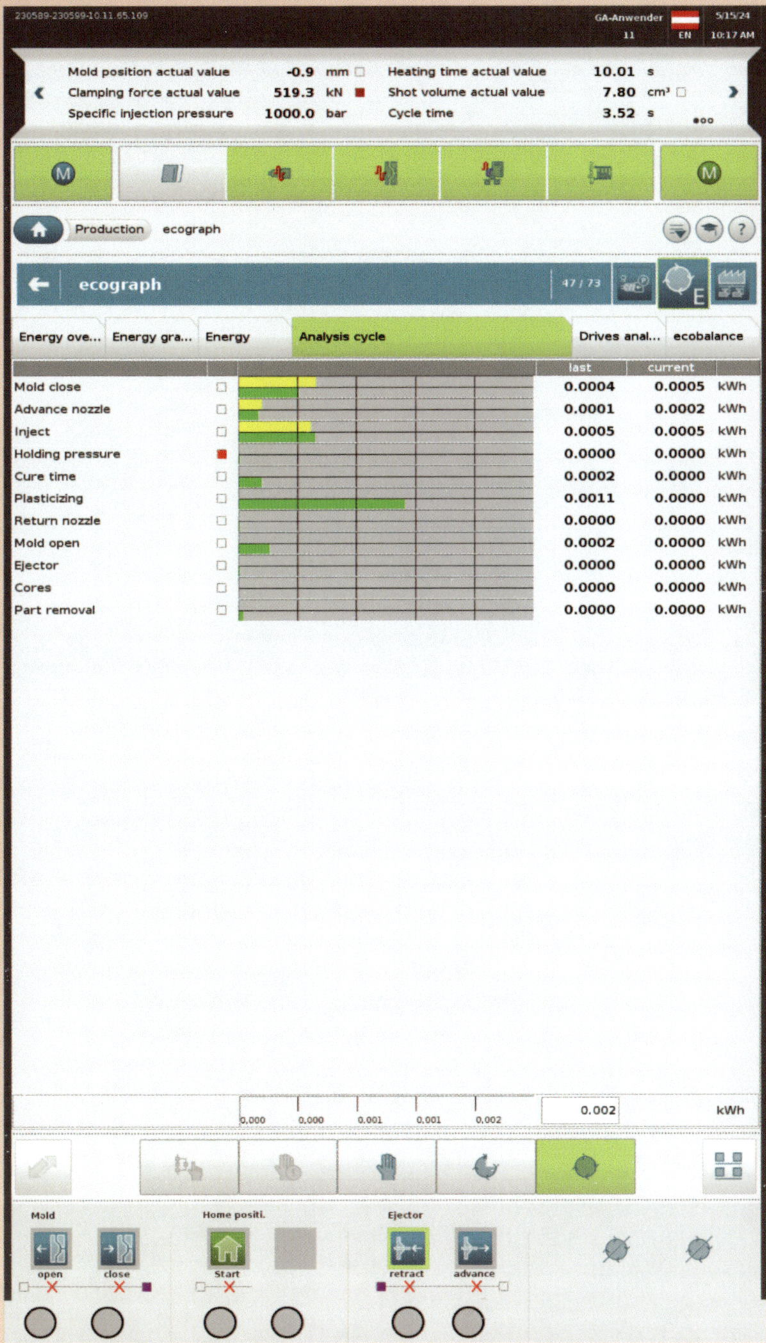

Figure 8.17 Energy monitoring in the machine controller (image source: Ecograph, Engel GmbH (*continued*)

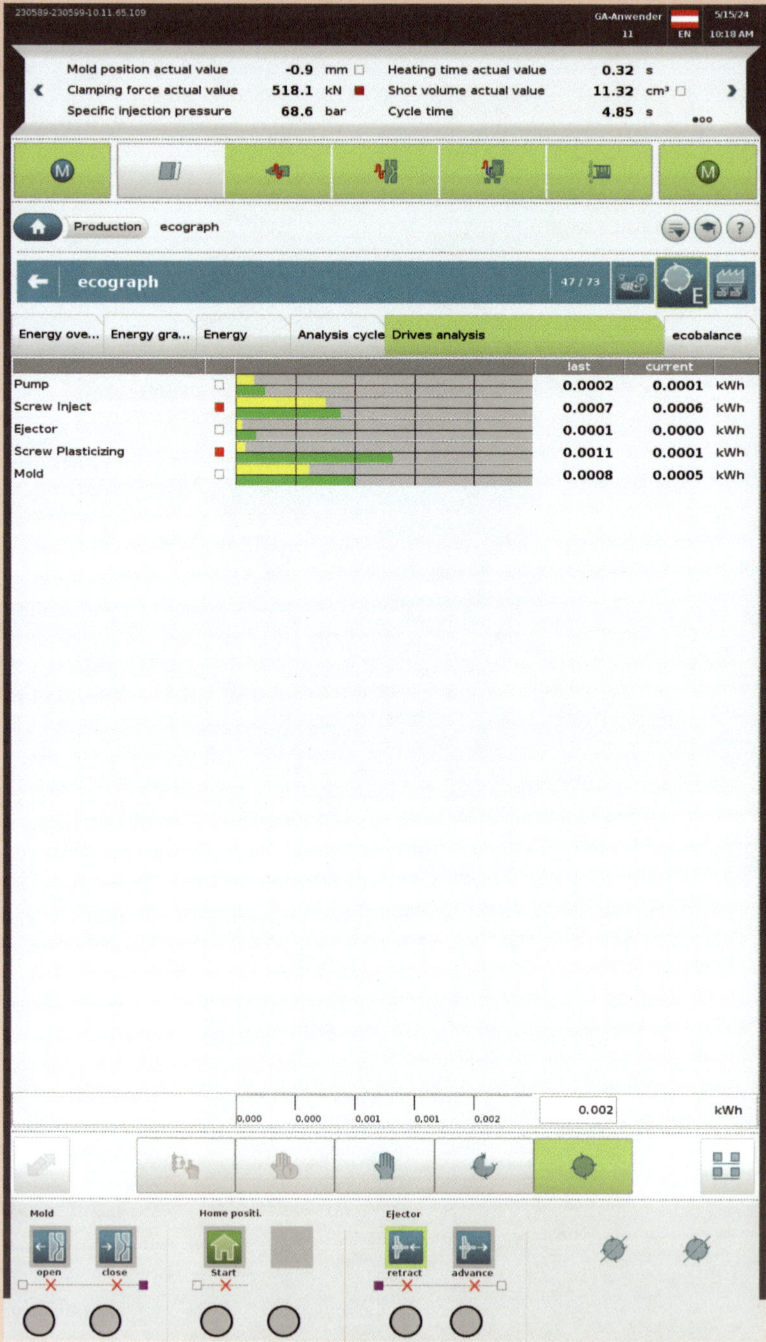

Figure 8.17 Energy monitoring in the machine controller (image source: Ecograph, Engel GmbH (*continued*)

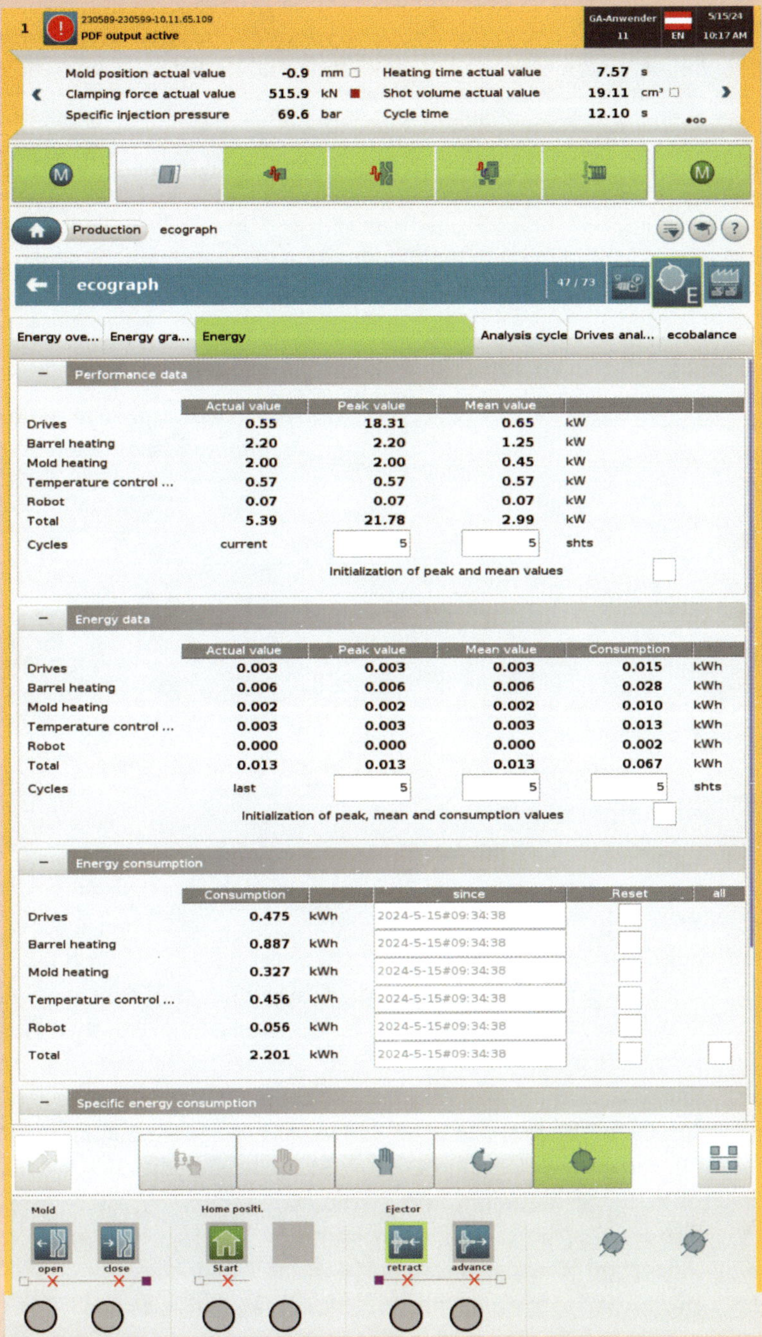

Figure 8.17 Energy monitoring in the machine controller (image source: Ecograph, Engel GmbH (*continued*)

For older machines, it is possible to have energy monitoring retrofitted by the machine manufacturer or by independent suppliers. In this case, a measuring module is permanently integrated into the injection molding machine once. This allows the power consumption for heaters, drives, etc. to be read on a separate monitor on the machine.

It is also possible to implement energy monitoring directly on the machine by using a power meter to make a manual measurement (Figure 8.18). This form of measurement is suitable for optimization during sampling, but should be used as a last resort for energy monitoring on the injection molding machine.

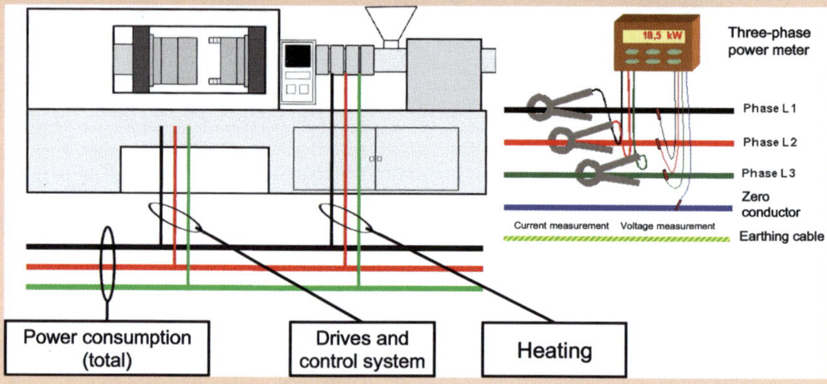

Figure 8.18 Measurement directly via hand-held measuring device

IMPORTANT WARNING:

For safety reasons, direct measurement with a hand-held measuring device in the machine's control cabinet may only be performed by qualified electricians.

Once the actual energy consumption (kWh) of the injection molding machine has been recorded and documented via energy monitoring, the next step is to evaluate and classify the actual consumption of the machine in terms of its energy efficiency. This is the only way to make further targeted decisions as to whether an improvement in the energy efficiency of the optimized basic setting is necessary. If only the actual energy consumption of the injection molding machine is evaluated, influencing factors of the machine setting (e.g. process parameters) will not be taken into account. In order to gain more meaningful knowledge, the specific energy consumption (kWh/kg) must be used as a comparative parameter. As already briefly presented in Chapter 1 (Section 1.4.4 "Specific energy consumption"), the specific energy consumption (kWh/kg) is the ratio of measured active power (kWh/kg) to processed plastic (kg). Consequently, in addition to the measured actual energy consumption of the machine, the shot weight (g) and the cycle time (s) are also taken into account. This results in an application-related value for the energy requirement of the optimized basic setting for the "mold sampling specialist".

In order to be able to compare energy differences between different process settings or injection molding machines, a specific energy consumption (kWh/kg) must generally be used. Only in this way can energy values be considered realistically and practically.

If the injection molding machine is working at high capacity and therefore has a high material throughput, energy efficiency can be improved. The reason for this is that the existing power loss on the machine is distributed over a larger quantity of plastic to be processed, regardless of the material throughput. Practical experience shows that very fast processes, such as in the packaging industry, have a fundamentally lower specific energy consumption than technical injection molded parts. This is usually due to the complex dimensions of technical molded parts, higher values of machine setting parameters, higher wall thicknesses and a longer required residual cooling time and thus a longer cycle time.

8.2.2.1 Evaluate and Classify Optimized Settings for Energy Efficiency

Following the energy monitoring, the specific energy consumption (kWh/kg) of your existing optimized basic setting must be determined. The specific energy consumption (kWh/kg) can be determined from the following actual values:

- Shot weight (g)
- Cycle time (s)
- Actual energy consumption (active power) of the injection molding machine per hour (kWh/h)

If all the required actual values are available, you can calculate the specific energy consumption (kWh/kg) of your machine or the optimized basic setting in two steps as follows:

Step 1: Material throughput per hour

$$\text{Throughput per hour}\left(\frac{\text{kg}}{\text{h}}\right)=\frac{3600\ \text{s}\times\text{Shot weight (g)}}{\text{Cycle time (s)}\times 1000} \tag{8.1}$$

Step 2: Specific energy consumption

$$\text{Spec.energy consumption}\left(\frac{\text{kWh}}{\text{kg}}\right)=\frac{\text{Actual energy consumption}\left(\frac{\text{kWh}}{\text{h}}\right)}{\text{Throughput per hour}\left(\frac{\text{kg}}{\text{h}}\right)} \tag{8.2}$$

Once the specific energy consumption (kWh/kg) of the optimized basic setting has been determined, this consumption must be evaluated. This evaluation is carried out using a classification (Table 8.3).

Table 8.3 Classification of Specific Energy Consumption (kWh/kg)

Valuation	Specific energy consumption	Classification
	< 0.50 kWh/kg	Very good
	0.50–0.75 kWh/kg	Good
	0.75–1.00 kWh/kg	Satisfactory → Optimization possible
	> 1.0 kWh/kg	Inadequate → Optimization necessary

In the following information box, I would like to calculate the specific energy consumption for the optimized basic setting from our case study of the housing cover.

PRACTICAL EXAMPLE:

GIVEN:

- Molded part: Housing cover
- Plastic: PC/ABS
- Shot weight: 260 g
- Cycle time: 50 s

Sought:

Specific energy consumption (kWh/kg)

Calculation:

Measured:

Actual energy consumption (active power) in kWh/h: 15.9

$$\text{Throughput per hour}\left(\frac{\text{kg}}{\text{h}}\right) = \frac{3600\ \text{s} \times 260\ \text{g}}{50\ \text{s} \times 1000} = 18.72\ \frac{\text{kg}}{\text{h}}$$

$$\text{Spec.energy consumption}\left(\frac{\text{kWh}}{\text{kg}}\right) = \frac{15.9\ \left(\frac{\text{kWh}}{\text{h}}\right)}{18.72\ \left(\frac{\text{kg}}{\text{h}}\right)} = 0.84\ \frac{\text{kWh}}{\text{kg}}$$

Following the evaluation and classification of your current specific energy consumption of the optimized basic setting, the next step is to make it more energy-efficient. In the following section, you will see which machine setting parameters you can use to achieve an increase in energy efficiency and an optimum approach to optimization.

REMEMBER:

The aim of any optimization should therefore be the following: Consistent quality of the molded part with the same productivity and significantly lower energy consumption.

8.2.3 Optimization Options for Increasing Energy Efficiency

If the specific energy consumption of the optimized basic setting is not optimal, this can be improved with the help of specially selected machine setting parameters. Table 8.4 shows an overview of all relevant machine setting parameters along with an assessment of their influence on energy consumption.

Table 8.4 Influence of Machine Parameters on Energy Consumption

Barrel temperature	++
Plasticizing	++
Holding pressure and holding pressure time	–
Residual cooling time	++
Clamping force	+
Mold movements	–
Mold temperature	++
Ejection	°
Movement (plasticizing unit)	°
Valuation:	
++ = Very large influence	– = Small influence
+ = Large influence	° = No influence

Numerous practical studies have shown that the following machine setting parameters can have an influence on energy consumption:

- Barrel temperature (°C)
- Mold temperature (°C)
- Plasticizing speed (1/min)
- Residual cooling time (s)
- Mold clamping force (kN)

Optimization of energy efficiency

Article description:	
Mold number:	
Mold sampling specialist:	
Date:	

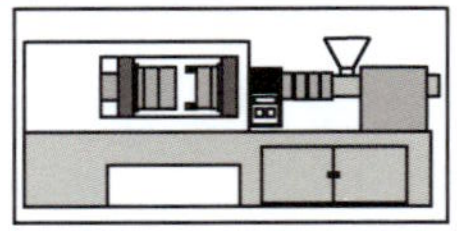

Machine setting / Evaluation method	Optimized basic setting			Injection molding test 1			Injection molding test 2			....
	Barrel temperature	value:		Barrel temperature	value:		Barrel temperature	value:		
	Mold temperature	value:		Mold temperature	value:		Mold temperature	value:		
	Plasticizing speed	value:		Plasticizing speed	value:		Plasticizing speed	value:		
	Residual cooling time	value:		Residual cooling time	value:		Residual cooling time	value:		
	Mold clamping force	value:		Mold clamping force	value:		Mold clamping force	value:		
Energy monitoring:	Actual energy consumption	kW/h		Actual energy consumption	kW/h		Actual energy consumption	kW/h		
	Specific energy consumption	kWh/kg		Specific energy consumption	kWh/kg		Specific energy consumption	kWh/kg		

Measurement	Direction	Nominal dimension	Tolerance	Part 1	Part 2	Part 3	Part 4	Part 5	Part 6	Mean value	Part 1	Part 2	Part 3	Part 4	Part 5	Part 6	Mean value	Part 1	Part 2	Part 3	Part 4	Part 5	Part 6	Mean value
	X-Direction																							
	Y-Direction	Nominal dimension	Tolerance	Part 1	Part 2	Part 3	Part 4	Part 5	Part 6	Mean value	Part 1	Part 2	Part 3	Part 4	Part 5	Part 6	Mean value	Part 1	Part 2	Part 3	Part 4	Part 5	Part 6	Mean value
	Z-Direction	Nominal dimension	Tolerance	Part 1	Part 2	Part 3	Part 4	Part 5	Part 6	Mean value	Part 1	Part 2	Part 3	Part 4	Part 5	Part 6	Mean value	Part 1	Part 2	Part 3	Part 4	Part 5	Part 6	Mean value

Visual assessment	Note:	Part 1	Part 2	Part 3	Part 4	Part 5	Part 6	Average result	Part 1	Part 2	Part 3	Part 4	Part 5	Part 6	Average result	Part 1	Part 2	Part 3	Part 4	Part 5	Part 6	Average result
	● Deterioration ● Neutral - No change																					

Figure 8.19 The "Optimization of energy efficiency" template

To enable a structured procedure, I have also created a template "Optimization of energy efficiency" (Figure 8.19). You can use this template to record or document the injection molding tests for "Optimizing energy efficiency" and make a fast comparison between the energy consumption tests.

As an increase in energy efficiency must never be achieved at the expense of the quality of the molded part or the productivity of the injection molding process, it is very important to conduct a dimensional accuracy and visual inspection of the molded part in parallel during optimization tests of the selected machine setting parameters on the machine. In this way, you can ensure that optimization to increase energy efficiency does not have a negative impact on your already optimized basic settings.

Using the housing cover as a case study from practice, I would now like to show you the individual steps of how energy efficiency optimization is implemented with the template.

PRACTICAL EXAMPLE:

Now remove the template (Figure 8.20) and enter the general data ① at the top of the header. Next, enter the selected machine setting parameters from the already optimized setting ②. Then, enter the actual energy consumption (kW/h) determined via energy monitoring and the calculated specific energy consumption (kWh/kg) under the "Optimized basic setting" field ③.

Then transfer the materials ④ from the "Evaluation of injection molding tests" template (Figure 8.10). Now, implement injection molding tests and enter the values of the machine setting parameters shown under the respective injection molding test ⑤. After the tests, compare the specific energy consumptions with each other. If an improvement in energy consumption was achieved, this setting must be examined and evaluated for dimensional accuracy and visual defects of the molded part ⑥. The practical example shows that a reduction in barrel and mold temperatures resulted in a deterioration of the molded part surface. Furthermore, a possible reduction in residual cooling time (with the help of thermal images) and a slower plasticizing speed were investigated. As a result of these tests, a satisfactory specific energy consumption of approx. 0.84 kWh/kg has already been reduced to a good specific energy consumption of approx. 0.75 kWhkg while maintaining the same molded part quality.

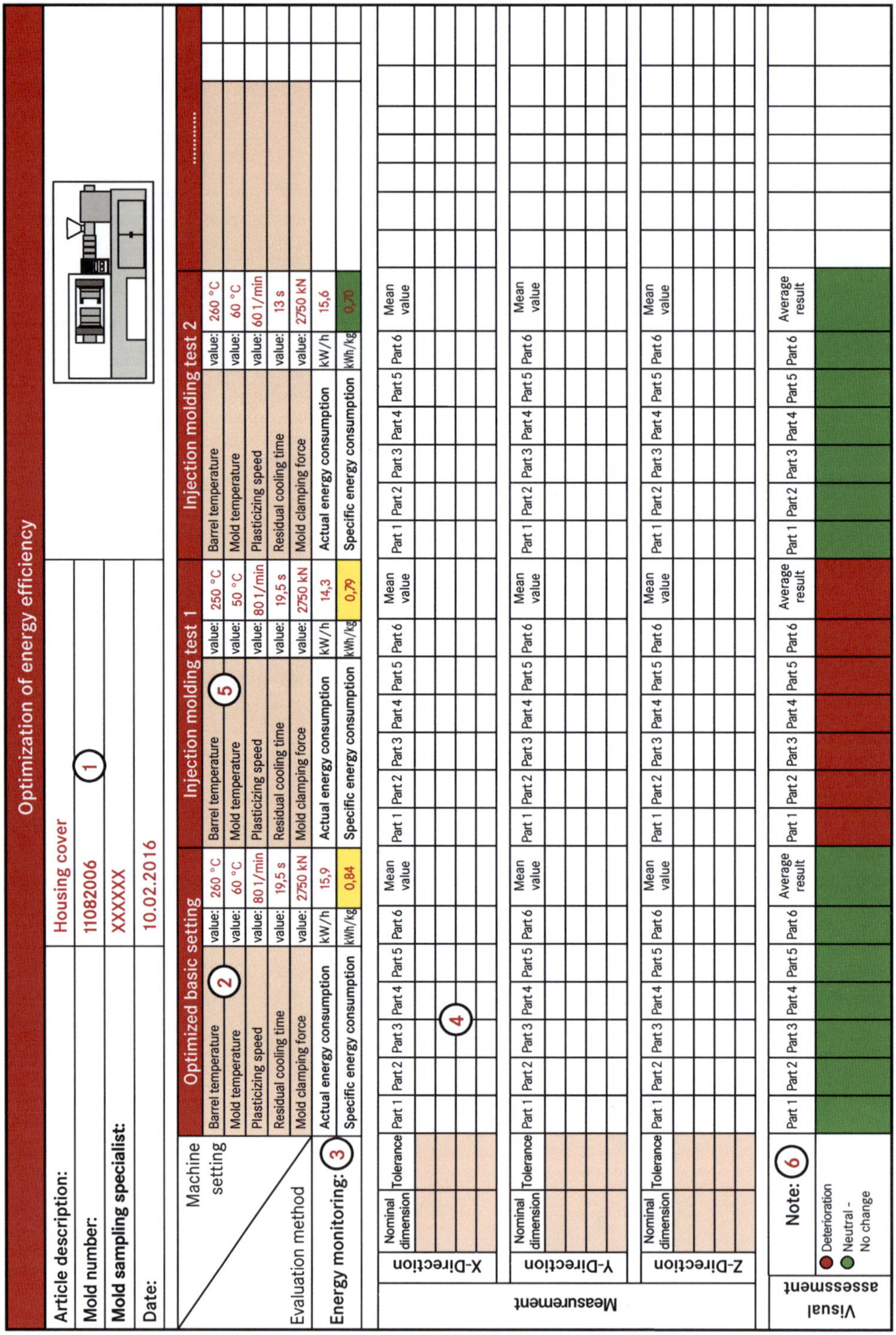

Figure 8.20 Practical example: Optimization of energy efficiency

8.3 Optimizing the Basic Setting—Part 3

Once you have worked through the optimization of your basic settings with regard to faults occurring on the molded part, productivity and energy efficiency in a structured manner, the final step is to implement the process capability of the optimized basic settings and a process Run@Rate. In the final sections of this chapter, I would therefore like to show you why a process capability analysis and a process Run@Rate are so important. The approach is shown in detail for the case of our housing cover and with the aid of practical examples.

8.3.1 Step 1: Process Capability Analysis of Machine Setting and Process Parameters

Once an optimized basic setting has been found, the extent to which the individual machine setting parameters deviate from the target value in the injection molding process must be investigated.

Consequently, the relevant machine setting parameters must now be recorded during the running injection molding process and evaluated using a capability analysis. The process capability analysis is divided into two areas called process capability (c_{pk}) and machine capability (c_{mk}).

The difference between the two capability analyses is that random samples of a process capability (c_{pk}) from a longer continuous production period at defined intervals serve as the basis for the capability calculation. From the process capability, it can be seen whether certain trends become visible within a period of time. In contrast, for machine capability (c_{mk}), a production period is selected by generally taking a number (approx. 50 parts) of molded parts produced in succession and using them as the basis for the capability calculation. The machine capability can be used to estimate whether the injection molding machine with its set machine setting parameters is at all capable of ensuring a reproducible and constant subsequent production process.

In the plastics processing industry, the process capability (c_{pk}) of the molded parts is generally determined and evaluated. This usually involves quality characteristics, such as dimensional accuracy or the weight of the molded part. As a result, it is possible to determine whether the molded parts were manufactured using a consistent injection molding process. As a process and machine capability analysis is dealt with specifically in numerous good text books, I will not go into this in detail here.

Instead, I would like to show the "mold sampling specialist" a method for checking the process reliability of the optimized basic setting on the injection molding machine. To this end, a capability test is performed to examine the fluctuation range of the machine setting and process parameters to be kept constant during production. For specific results for the fluctuation range of the machine setting and process parameters during sampling, I would like to explain a procedure developed in practice for capability evaluation on the machine. You can use it to analyze the set parameters for possible deviations from the target value, taking recommended tolerances into account, and thus determine and assess the capability or process reliability. As with the process or machine capability analysis, a target value with upper specification limit (USL = target value + upper tolerance limit) and lower specification limit (LSL = target value + lower tolerance limit) must first be known for this method. Table 8.5 shows the recommended tolerances for deviations from the target value for all important machine setting and process parameters. Depending on the molded part, an optional decision can be made here as to whether it is a precision, technical or general molded part.

Table 8.5 Recommended Empirical Tolerances for Machine Setting and Process Parameters

Tolerances for machine setting and process parameters:			
Machine setting and process parameters:	Tolerances in ±% of the setpoint		
	Precision parts	Technical parts	General parts
Injection pressure	± 4%	± 7%	± 10%
Plasticizing time	± 2%	± 4%	± 8%
Injection time	± 2%	± 4%	± 6%
Residual melt cushion	± 2%	± 5%	± 10%
Cycle time	± 2%	± 3%	± 4%
Specific cavity pressure	± 3%	± 4%	± 6%
Molded part weight	± 0.05%–0.10%	± 0.10%–0.30%	± 0.30%–0.50%

Using these recommended tolerances, you can now analyze all set values of the machine setting and process parameters shown for their fluctuation range and thus for their process capability. A "Process capability analysis of machine setting and process parameters" template (Figure 8.21) has been created for this purpose, in which the already optimized basic setting can be checked for its process capability. If it is process-capable and the deviations or fluctuation range are within the recommended tolerance, these tolerance values can then be used for quality monitoring in the injection molding machine.

Process capability analysis of machine setting and process parameters			
Article description:			
Mold number:			
Mold sampling specialist:			
Date:			

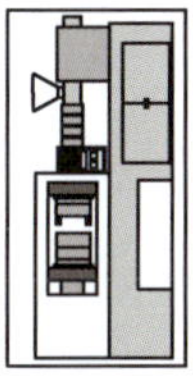

Recommended tolerance range of the set machine setting and process parameters in %:			
Parameters	**Precision parts**	**Technical parts**	**Technical parts**
Injection pressure	±4 %	±7 %	±10 %
Plasticizing time	±2 %	±4 %	± 8 %
Injection time	±2 %	±4 %	± 6 %
Residual melt cushion	±2 %	±5 %	±10 %
Cycle time	±2 %	±3 %	± 4 %
Spec. cavity pressure	±3 %	±4 %	± 6 %
Molded part weight	±0,05 % - 0,10 %	±0,10 % - 0,30 %	±0,30 % - 0,50 %

Evaluation of process capability:

Parameters	Number of molded parts	Setpoint	Mean value of the actual state	Variation range in %	Tolerance in %	ok	not ok
Injection pressure							
Plasticizing time							
Injection time							
Residual melt cushion							
Zykluszeit							
Cycle time							
Molded part weight							

Figure 8.21 The "Process capability analysis of machine setting and process parameters" template

The following information box explains the process capability analysis procedure in detail using our case study of the housing cover.

PRACTICAL EXAMPLE:

Remove the "Process capability analysis of machine setting and process parameters" template and enter the general data ①. The next step is to decide what type of molded part it is, e. g. precision, technical or general molded part ②. Our example is a technical molded part. Now, produce 40–50 molded parts with your "optimized basic setting" and take the individual values from the injection molding machine and enter them under the listed machine setting and process parameters as well as the molded part weight in the respective boxes ③.

PRACTICAL TIP:

Depending on the make of your injection molding machine, you can use your quality monitoring table (Figure 8.22) in the machine control system to record and transfer the relevant machine setting and process parameters, such as injection pressure, plasticizing time, injection time, residual melt cushion, cycle time and cavity pressure of the 40–50 molded parts produced.

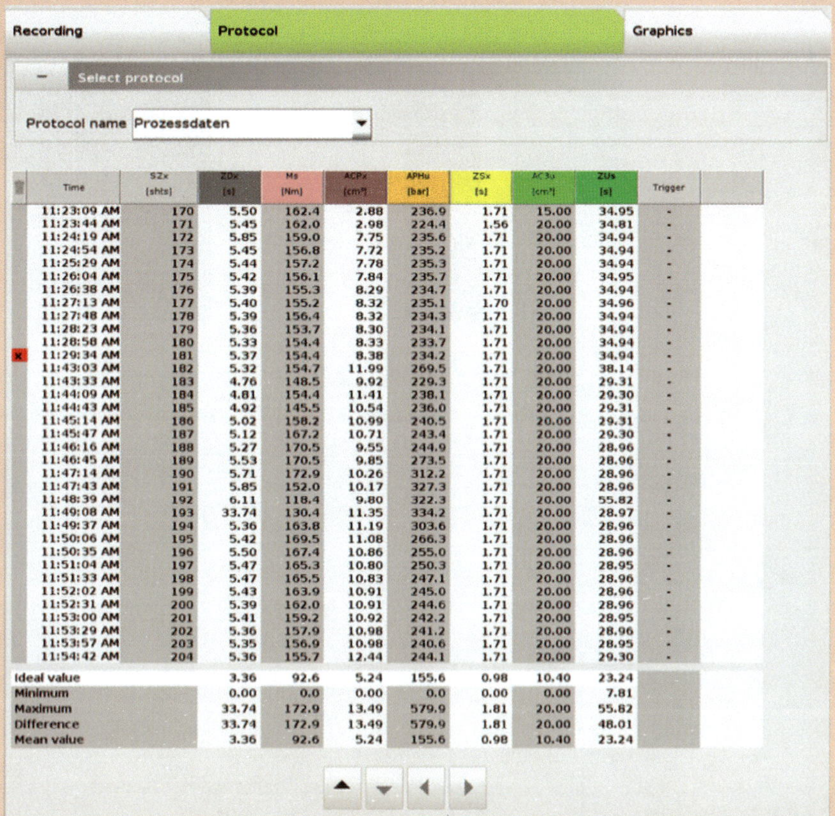

Recording | Protocol | Graphics

Select protocol

Protocol name: Prozessdaten

Time	SZx [shts]	ZDx [s]	Ms [Nm]	ACPx [cm³]	APHu [bar]	ZSx [s]	AC3u [cm³]	ZUs [s]	Trigger
11:23:09 AM	170	5.50	162.4	2.88	236.9	1.71	15.00	34.95	-
11:23:44 AM	171	5.45	162.0	2.98	224.4	1.56	20.00	34.81	-
11:24:19 AM	172	5.85	159.0	7.75	235.6	1.71	20.00	34.94	-
11:24:54 AM	173	5.45	156.8	7.72	235.2	1.71	20.00	34.94	-
11:25:29 AM	174	5.44	157.2	7.78	235.3	1.71	20.00	34.94	-
11:26:04 AM	175	5.42	156.1	7.84	235.7	1.71	20.00	34.95	-
11:26:38 AM	176	5.39	155.3	8.29	234.7	1.71	20.00	34.94	-
11:27:13 AM	177	5.40	155.2	8.32	235.1	1.70	20.00	34.96	-
11:27:48 AM	178	5.39	156.4	8.32	234.3	1.71	20.00	34.94	-
11:28:23 AM	179	5.36	153.7	8.30	234.1	1.71	20.00	34.94	-
11:28:58 AM	180	5.33	154.4	8.33	233.7	1.71	20.00	34.94	-
11:29:34 AM	181	5.37	154.4	8.38	234.2	1.71	20.00	34.94	-
11:43:03 AM	182	5.32	154.7	11.99	269.5	1.71	20.00	38.14	-
11:43:33 AM	183	4.76	148.5	9.92	229.3	1.71	20.00	29.31	-
11:44:09 AM	184	4.81	154.4	11.41	238.1	1.71	20.00	29.30	-
11:44:43 AM	185	4.92	145.5	10.54	236.0	1.71	20.00	29.31	-
11:45:14 AM	186	5.02	158.2	10.99	240.5	1.71	20.00	29.31	-
11:45:47 AM	187	5.12	167.2	10.71	243.4	1.71	20.00	29.30	-
11:46:16 AM	188	5.27	170.5	9.55	244.9	1.71	20.00	28.96	-
11:46:45 AM	189	5.53	170.5	9.85	273.5	1.71	20.00	28.96	-
11:47:14 AM	190	5.71	172.9	10.26	312.2	1.71	20.00	28.96	-
11:47:43 AM	191	5.85	152.0	10.17	327.3	1.71	20.00	28.96	-
11:48:39 AM	192	6.11	118.4	9.80	322.3	1.71	20.00	55.82	-
11:49:08 AM	193	33.74	130.4	11.35	334.2	1.71	20.00	28.97	-
11:49:37 AM	194	5.36	163.8	11.19	303.6	1.71	20.00	28.96	-
11:50:06 AM	195	5.42	169.5	11.08	266.3	1.71	20.00	28.96	-
11:50:35 AM	196	5.50	167.4	10.86	255.0	1.71	20.00	28.96	-
11:51:04 AM	197	5.47	165.3	10.80	250.3	1.71	20.00	28.95	-
11:51:33 AM	198	5.47	165.5	10.83	247.1	1.71	20.00	28.96	-
11:52:02 AM	199	5.43	163.9	10.91	245.0	1.71	20.00	28.96	-
11:52:31 AM	200	5.39	162.0	10.91	244.6	1.71	20.00	28.96	-
11:53:00 AM	201	5.41	159.2	10.92	242.2	1.71	20.00	28.95	-
11:53:29 AM	202	5.36	157.9	10.98	241.2	1.71	20.00	28.96	-
11:53:57 AM	203	5.35	156.9	10.98	240.6	1.71	20.00	28.95	-
11:54:42 AM	204	5.36	155.7	12.44	244.1	1.71	20.00	29.30	-
Ideal value		3.36	92.6	5.24	155.6	0.98	10.40	23.24	
Minimum		0.00	0.0	0.00	0.0	0.00	0.00	7.81	
Maximum		33.74	172.9	13.49	579.9	1.81	20.00	55.82	
Difference		33.74	172.9	13.49	579.9	1.81	20.00	48.01	
Mean value		3.36	92.6	5.24	155.6	0.98	10.40	23.24	

Figure 8.22 Example of a quality table for an injection molding machine

Once all values have been entered in the template, enter the target values for the individual machine setting and process parameters ④. The next step is to calculate the average value of the individual parameters as follows ⑤:

$$\text{Mean value } x = \frac{\text{Sum of all recorded measured values}}{\text{Number of measured values recorded}}$$

Once the mean value has been determined, the next step is to enter the variation range between the minimum and maximum measured value of the respective parameter in the template ⑥. The next step is to calculate the variation range from the target value as a percentage ⑦, using the following equation:

$$\text{Variation } (\%) = \frac{\text{Total variation range } (\pm)}{\text{Mean value } x} \times 100$$

Finally, the right-hand column is used to check ⑦ whether the specified recommended tolerance has been met. If this is the case, process capability or process reliability for subsequent series production has been achieved.

Process capability analysis of machine setting and process parameters

Article description:	Gehäuseabdeckung	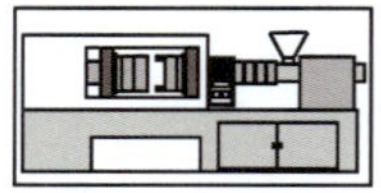
Mold number:	11082006 (1)	
Mold sampling specialist:	xxxxxxxx	
Date:	16.01.2018	

Recommended tolerance range of the set machine setting and process parameters in %:

Parameters	Precision parts	Technical parts (2)	Technical parts
Injection pressure	±4 %	±7 %	±10 %
Plasticizing time	±2 %	±4 %	± 8 %
Injection time	±2 %	±4 %	± 6 %
Residual melt cushion	±2 %	±5 %	±10 %
Cycle time	±2 %	±3 %	± 4 %
Spec. cavity pressure	±3 %	±4 %	± 6 %
Molded part weight	±0.05 % – 0.10 %	±0.10 % – 0.30 %	±0.30 % – 0.50 %

Evaluation of process capability: (4) (5) (6) (7) (8)

Parameters	Number of molded parts																				Setpoint	Mean value of the actual state	Variation range in %	Tolerance in %	ok	not ok
Injection pressure	980	990	985	980	988	983	979	985	980	980	984	980	985	990	980	983	983	980	980	982	980.00	983.55	Max. 990.00	11.00	±7 %	X
	980	980	984	980	985	990	990	985	980	988	980	988	983	980	990	985	980	988	983	979			Min. 979.00			
Plasticizing time	6.50	6.51	6.50	6.50	6.52	6.50	6.49	6.51	6.50	6.50	6.50	6.53	6.49	6.52	6.50	6.51	6.52	6.50	6.48	6.50	6.50	6.50	Max. 6.53	0.76	±4 %	X
	6.49	6.51	6.50	6.50	6.50	6.53	6.49	6.50	6.51	6.50	6.50	6.50	6.48	6.50	6.52	6.50	6.49	6.51	6.50	6.48			Min. 6.48			
Injection time	2.40	2.35	2.30	2.33	2.35	2.30	2.40	[illegible] (3)	2.30	2.34	2.33	2.30	2.39	2.33	2.30	2.40	2.33	2.33	2.30	2.30	2.30	2.33	Max. 2.40	4.29	±4 %	X
	2.40	2.35	2.30	2.32	2.35	2.31	2.33	[illegible]	2.33	2.35	2.30	2.30	2.33	2.30	2.33	2.30	2.40	2.33	2.33	2.33			Min. 2.30			
Residual melt cushion	8.54	8.55	8.60	8.50	8.53	8.51	8.45	8.50	8.50	8.49	8.53	8.54	8.57	8.50	8.48	8.52	8.51	8.54	8.59	8.55	8.50	8.50	Max. 8.60	1.76	±5 %	X
	8.54	8.55	8.60	8.50	8.53	8.51	8.57	8.50	8.48	8.52	8.51	8.54	8.55	8.60	8.50	8.53	8.50	8.49	8.53	8.51			Min. 8.45			
Zykluszeit	45	45	45.1	45	45.1	45	45	44.9	45.1	45.2	45.05	45	45	45	45.1	45.1	45	45.1	45	45.1	42.50	45.06	Max. 45.20	2.38	±3 %	X
	45	45	45.2	45.05	45	45	45	45.1	45.1	45.2	45	45.1	45	45.1	45.2	45.1	45	45	45.2	45.1			Min. 44.89			
Cycle time	440	445	453	455	451	450	445	550	453	457	453	453	450	454	445	450	450	448	447	450	450.00	452.35	Max. 550.00	2.21	±4 %	X
	450	448	447	450	453	453	450	440	445	453	455	451	450	453	453	440	445	453	455	451			Min. 440.00			
Molded part weight	261	260	260	261	261	262	261	260	261	260	260	261	261	260	260	261	260	260	261	260	260.00	260.50	Max. 261.50	0.57	±0.10–0.30 %	X
	261	261	260	261	260	260	260	260	261	260	261	260	261	260	260	260	261	262	260	261			Min. 260.00			

Figure 8.23 Practical example: Process capability analysis of machine setting and process parameters

8.3.2 Step 2: Process Run@Rate of the Optimized Basic Setting

In practice, it is unfortunately very often the case that actual cycle times, actual molded part weights and actual quantities with regard to rejects and good parts, for example, do not match or are not compared with the calculated target values. In most cases, these imputed target values originate from a quotation submitted to the customer by the sales department on the basis of a request for quotation, for example. These target values such as cycle time, weight of the molded part, plastic, target quantities with regard to rejects and good parts, size of the injection molding machine are important factors that determine the price of the molded part. As these imputed target values are purely theoretical estimates, it is particularly important to implement an actual/target value comparison of your optimized basic settings using a Process Run@Rate. Firstly, I would like to discuss the term "Process Run@Rate" in more detail.

BACKGROUND KNOWLEDGE:

What is Process Run@Rate?

Process Run@Rate originally comes from the automotive industry and is a performance test that the customer of the molded parts (OEM) conducts at the supplier. Process Run@Rate means that a production run is carried out over a certain period of time or over a certain quantity and it is then assessed whether the total cycle time and actual quantities with regard to rejects and good parts correspond to the calculated values. The aim is to check the production capability and quality of the molded part as well as the confirmation of the achievement of the series cycle times (capacity confirmation) before the start of series production.

The basic idea of a Process Run@Rate should generally be carried out internally in your company to review your own injection molding processes, regardless of whether you are in the automotive, medical, packaging or electrical industry. This is the only way to check whether the optimized basic setting corresponds to the calculated target values or whether a recalculation is required, for example by the sales department. For a Process Run@Rate, a template (Figure 8.24) has been created specifically for mold sampling, which the "mold sampling specialist" and the project manager can use to check, evaluate and document all important target/actual value comparisons fast and efficiently.

Result and summary (project manager/sales):	
Target:	**Calculated output rate**
Pieces/hour:	
Pieces/shift:	
Calculated Run@Rate duration (hours or number of parts)	
Calculated downtimes (hours):	
Reasons for calculated downtimes (hours):	
Molded part weight (g):	
Results:	**Actual output rate according to Run@Rate**
Pieces/hour:	
Pieces/shift:	
Actual Run@Rate duration (hours or number of parts)	
Actual downtimes (hours):	
Explanation for actual downtimes (hours):	
Molded part weight (g):	
Total number of units produced:	
Total scrap (units):	
Scrap rate in %:	
Comment:	
Evaluation (person responsible for sampling/distribution):	
passed ☐ Passed with conditions ☐	
not passed ☐ Recalculation required ☐	

Figure 8.24 The "Process Run@Rate of the optimized basic setting" template

To conclude the optimization of the basic setting, I will explain the procedure for a process Run@Rate and the handling of the example template in individual steps using our case study of the housing cover.

PRACTICAL EXAMPLE:

The "mold sampling specialist" takes the template (Figure 8.25) from the molding documents and fills in the top header ①. Depending on the requirements, whether internal or from the customer, a process Run@Rate with the optimized basic setting is implemented in a defined production time (hours) or a defined number of molded parts. Now, start the Run@Rate on the machine by running the optimized basic setting under series conditions. During the process Run@Rate, fill in the fields under the "Process monitoring" column ② in the template. At the end of the process Run@Rate, the template is passed on to the project manager or to sales. The specialist responsible then fills in the other points under the "Process analysis" column (project manager/sales) ③. Once all the important data has been collected, the optimized machine settings must now be evaluated ④. If the process Run@Rate is passed, the machine setting can be adopted as a series setting. If the actual values at the end of the process Run@Rate are higher than the calculated values, a recalculation must be carried out for economic reasons.

Result and summary (project manager/sales):	
Target:	**Calculated output rate**
Pieces/hour:	50
Pieces/shift:	72
Calculated Run@Rate duration (hours or number of parts)	④ 576
Calculated downtimes (hours):	
Reasons for calculated downtimes (hours):	
Molded part weight (g):	260
Results:	**Actual output rate according to Run@Rate**
Pieces/hour:	45
Pieces/shift:	80
Actual Run@Rate duration (hours or number of parts)	640
Actual downtimes (hours):	500
Explanation for actual downtimes (hours):	5
Molded part weight (g):	260,5
Total number of units produced:	500
Total scrap (units):	2
Scrap rate in %:	10
Comment:	Scrap due to machine restart after machine malfunction

⑤ Evaluation (person responsible for sampling/distribution):			
passed	[x]	Passed with conditions	[ ]
not passed	[ ]	Recalculation required	[ ]

Figure 8.25 Practical example: Process Run@Rate of the optimized basic setting

Milestone: "Optimization of the basic setting"

- Block 7 of the sampling checklist has been processed in full
- The preliminary basic setting of the molding process has been optimized for the following points:
  - Faults occurring on the molded part
  - Productivity
  - Energy efficiency
- The machine setting and process parameters have been analyzed for their process capability
- A process Run@Rate has been performed

9 Documentation of Mold Sampling

Following the optimization of the basic setting, it is now important to document all the results, knowledge gained, problems encountered etc. in the sampling process to date. In this chapter, I would like to show you why documentation is so important for the further optimization process. I will also show you how you can document not only all the important machine setting and process parameters on a machine setting data sheet, but also all the results, knowledge gained, problems encountered, etc. in the form of a mold sampling report. Documentation of mold sampling should generally be simple, occur in real time and be completed quickly. In the following sections, I would like to make you aware of the need for proper documentation during a mold sampling and explain how such documentation can or should be laid out.

The "mold sampling specialist" reviews block 9 ("Documentation of mold sampling") of the mold sampling checklist (Figure 9.1).

Documentation of mold sampling				
58	Machine setting and process parameters documented in the machine setting data sheet?			Block 9
59	All findings, problems, etc. from the mold inspection documented in detail under "Assessment by the mold sampling specialist" in the mold sampling report?			
60	All documents for the current mold sampling (templates) processed correctly and completely?			
61	Documentation handed over in full to the project manager?			

Figure 9.1 Mold sampling checklist: Block 9

9.1 Why Is Documentation So Important?

Well-managed documentation is very important for the subsequent sampling process in respect of being able to guarantee the following points:

- Traceability of the sampling process
- Checking that the correct procedure has been followed during the sampling process
- Secure knowledge transfer
- Collecting and archiving of empirical values
- Making problems and changes to the mold or injection molding process easy to see quickly for third parties, such as the designer, mold maker, project manager, customers
- Rendering the current status of mold sampling visible
- Complete collection of documents for subsequent hand-over to series production

9.2 Documentation of the Machine Setting and Process Parameters

Once an optimized machine setting has been found, it must not only be saved in the injection molding machine or on a data carrier, but also transferred in written form to a machine setting data sheet.

The machine setting data sheet (Figure 9.2) contains all the important machine setting and process parameters for the optimized basic setting. The data sheet has been divided into individual segments, such as

- General information
- Temperatures (machine setting parameters)
- Plasticizing unit (machine setting parameters)
- Clamping unit (machine setting parameters)
- Process parameters, and
- Energy efficiency analysis

The completed data sheet is attached to the injection molding machine at the end of the sampling process, together with the other sampling documents (templates) processed up to that point, as an appendix to the mold sampling report.

The machine setting data sheet can also be used as a possible basis for argumentation in the subsequent "Brief meeting (ideas-generation meeting) of all employees involved in the mold approval process" (Chapter 10) to define actions for dealing with problems that arise. This makes it possible for third parties to trace rapidly and in detail the optimized basic settings made by the "mold sampling specialist".

Process parameters					
Process times					
Residual cooling time:		s	Plasticizing time:		s
Cycle time:		s	Dwell time:		min
Break time:		s			
Process temperatures (manually Measured and checked)					
Mold nozzle side:			**Mold ejector side:**		
Mold wall temperature:		°C	Mold wall temperature:		°C
Mold flow temperature (temperature control unit):		°C	Mold flow temperature (temperature control unit):		°C
Mold return temperature (temperature control unit):		°C	Mold return temperature (temperature control unit):		°C
Melting temperature:		°C			
Demolding temperature of the molded part:		°C			
Process pressures:					
Required specific injection pressure:		bar	Cavity pressure (Max.):		bar
Process routes:					
Residual mass cushion:		❐ mm ❐cm³			

Energy efficiency analysis							
Specific energy consumption:							kWh/kg
< 0.5	very good	0.5 to 0.75	good	0.75 to 1.0	satisfactory	> 1.0	bad

Weight					
Molded part:		g	Sprue:		g
Total:		g			

Flow rate (temperature control medium)			
Measured flow rate:		L/min	

Figure 9.2 The "Machine setting data sheet" template

9.3 Mold Sampling Report

As already described, correct mold sampling documentation is essential for the continued success of the sampling process. In order to achieve effective, structured and transparent documentation, a mold sampling report (Figure 9.3 to Figure 9.7) has been designed specifically for this purpose. This sampling report is divided into individual sections according to the employees involved in the sampling and is labeled with the respective icons.

NOTE:

At the end of the sampling process, the aim must be to have a fully completed sampling report with all processed sampling documents (templates) in the appendix.

The "mold sampling specialist" begins to document all results, knowledge, problems etc. from the previous sampling process in real time during the sampling process in the section "Assessment by the mold sampling specialist" (Figure 9.3). When doing so, the specialist compares whether there are any deviations that have been checkmarked with a "No" on the "sampling checklist" en route. If there are deviations from the sampling checklist during sampling, these must be documented in the report. Furthermore, all problems that occur or are recognized during the sampling process must be properly recorded in writing. In order to ensure clarity and transparency, the section "Assessment by the mold sampling specialist" has been divided into process, part and mold (see Figure 9.3).

Mold sampling report					
Article description:			Bauteil:		
Article number:					
Mold number:					
Sample date:					
Injection molding machine:					
Customer:					
Type of sampling:	❒ Initial sampling (FOT)		❒ Material samples		
	❒ Sequential sampling (iteration loop)		❒ Process optimization (DoE)		
	❒ Mold change		❒ Customer complaint		
Employees involved in the sign-off:					
On the process (problems/comments):		Name:	Specialist for construction:	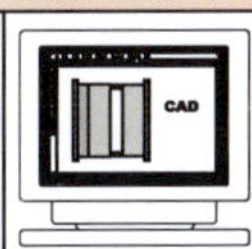	Name:
Machine operator:	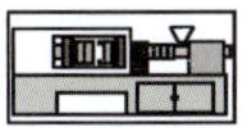	Name:	Specialist for quality management/ measurement technology:	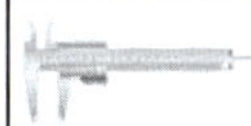	Name:
Specialist for mold making:	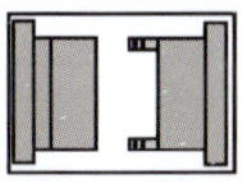	Name:			

Assessment by the machine operator for sampling:	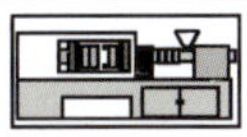
On the process (problems/comments):	
To the mold (problems/comments):	
To the molded part (problems/comments):	

Figure 9.3 Report section: "Assessment by the mold sampling specialist"

Assessment by the design engineer:	
(Problems/comments):	

Assessment by the quality assurance/metrology specialist:	
(Problems/comments):	

Figure 9.4 Report sections: "Assessments by the design and quality assurance specialists"

Assessment by the mold making specialist:	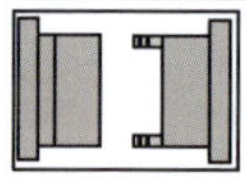
(Problems/comments):	

Assessment by the person in charge of the sampling:	
(Problems/comments):	

Figure 9.5 Report sections: "Assessment by the mold-making specialist and the project manager"

Determination of measures (tool corrections or changes, DoE etc.):	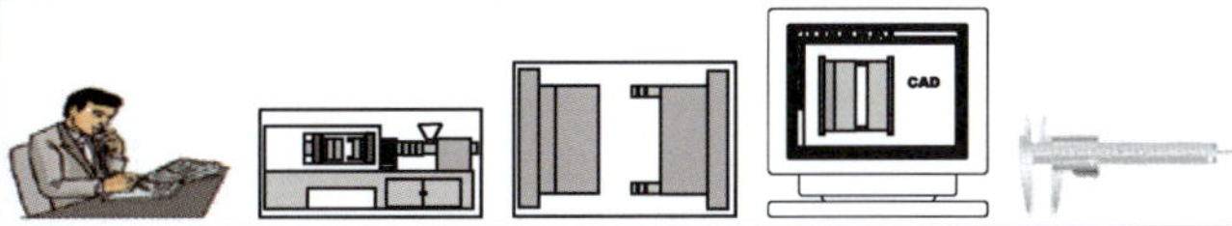

Further procedure:	
(Please tick the box):	
Release process:	❐ Yes ❐ no
Mold changes or corrections completed by:	
Preparation of a further subsequent sampling (iteration loop):	❐ Yes ❐ no

Figure 9.6 Report section: "Definition of action plan and subsequent procedure"

Check that all sign-off documents are complete:		
(Please tick the box):		
	okay	not okay
Information sheet for validations		
Sampling checklist		
Material data sheet on processing instructions from the material manufacturer		
Mold cooling plan		
Template "Water connection plan"		
Template "Mold sampling report"		
Template "Determining the required holding pressure time"		
Template "Experimental optimization of the mold clamping force"		
Template "Machine setting data sheet"		
Template "One-factor method/design of experiments (DoE)"		
Template "Evaluation of the injection molding tests"		
Template "Optimization of energy efficiency"		
Template "Process Run@Rate of the optimised basic setting"		
Template "Process capability analysis of machine setting and process parameters"		

Figure 9.7 Report section: "Checking the completeness of all documents"

If, in addition to the machine setting and process parameters, all results, knowledge, problems encountered etc. from the sampling process to date have been documented by the "mold sampling specialist", the initial sampling on the injection molding machine has been successfully completed. Following the subsequent brief meeting (ideas-generation meeting) of all employees involved in the sampling process (Chapter 10), the outstanding assessments and the jointly agreed actions as well as the subsequent procedure are documented in the mold sampling report.

9.4 Convene a Brief Meeting with All Employees/ Departments Involved in the Process

The "mold sampling specialist" informs the project manager when sampling on the injection molding machine is complete and hands over the "mold sampling report" along with all the documents.

Using the checklist at the end of the mold sampling report (Figure 9.7), the project manager checks the documents for completeness. If all documents have been filled out completely and correctly, a brief meeting (ideas-generation meeting) of all employees involved in the mold sampling process (Chapter 10) must be convened promptly to determine the actions and next steps to be taken.

Milestone: "Documentation of mold sampling"

- Block 8 of the "mold sampling checklist" has been processed
- Machine setting and process parameters are documented in the machine setting data sheet
- The "Assessment by the mold sampling specialist" section in the mold sampling report has been completed in full
- The assessments of the specialists for design, mold-making, quality assurance/metrology and the project manager are still outstanding (see Chapter 10 "Brief meeting and action plan")
- Sampling on the injection molding machine has been completed
- The mold sampling report has been handed over, along with all documents, and checked for completeness
- The brief meeting (ideas-generation meeting) of all employees involved in the mold sampling process has been convened promptly by the project manager

10 Brief Meeting and Action Plan

In plastics processing companies, there are major deficits in communication between the employees involved in the sampling process, in addition to a lack of documentation. This results in imprecise agreements between them, which means that the causes of problems, which form the basis for the action plan and the subsequent procedure, cannot be optimally identified.

In this chapter, I would like to introduce you to the brief meeting, which can significantly improve communication between all employees involved in the process. The advantages of a brief meeting and the procedure are shown. I will also show you how this can be realized fast and effectively in a very short time using targeted techniques to save resources.

10.1 Brief Meeting (Ideas-Generation Meeting)

10.1.1 Benefits of a Brief Meeting

The brief meetings held in practice to define actions have the following advantages:

- More targeted and effective teamwork with five areas of specialist expertise creates an optimal and more meaningful action plan.
- Time savings through finding a solution faster as a team (group dynamic).

Cost savings through well thought-out action plan in the team and thus avoidance of wrong decisions that lead to further unnecessary rejections (iteration loops).
A learning process (lessons learned) takes place when determining the cause of the problem, which enables errors to be detected earlier in other sampling processes, e. g. at an early stage in mold and part design or in mold making.
Improved communication between the departments involved in the sampling process, e. g. scheduling agreements, status of sampling, problems that arise, action plan
A transparent approach by the participants.
Improved cooperation within the company.

10.1.2 General Information on the Brief Meeting

Nowadays, companies schedule numerous meetings that focus on a wide variety of points, but all meetings have a common problem–the time that has to be invested in them.

What does often go fundamentally wrong at these meetings?

- There is no concrete objective or agenda
- Participants do not get to the heart of the matter and are not brief. They see the meeting as a place for self-promotion
- No communication rules are observed; you are interrupted by your discussion partners when speaking
- Some or all of the participants are not well prepared for the meeting and therefore do not understand what others are talking about
- The actual topic of the meeting is lost sight of, which means the meeting is no longer purposeful
- The meeting ends in an argument with mutual recriminations about faults and is therefore no longer efficient

This list could be continued indefinitely just for sampling meetings in a company. For economic reasons (resources), the action plan should therefore be implemented very efficiently and in a time-saving manner in the meeting. A duration of maximum 15 to 30 minutes has proven itself in practice across numerous samplings.

The following conditions must be met in order for this time period to be realized:

- Proper preparation for the brief meeting
- A neutral meeting location
- Creative methods for finding solutions faster

10.1.3 Proper Preparation for the Brief Meeting

The project manager is responsible for ensuring that all necessary documents are complete and correctly filled out for the brief meeting. All employees involved in the process must be informed of the process from the outset.

10.1.4 A Neutral Meeting Location

The brief meeting should always take place at the same location. It should be a neutral location and not, for example, the office of the project manager. The advantage of a neutral location is that the scheduled meetings quickly become an everyday routine procedure for the employees involved in the process. The meeting location should also fulfill other criteria, such as:

- Power connection for tablet, laptop, projector etc.
- Blackboard or whiteboard for sketches, notes, mind mapping

10.1.5 Creative Methods for Finding Solutions Faster

The methods of "brainstorming" and "mind mapping" make for a fast and efficient brief meeting.

10.1.5.1 Brainstorming

What is brainstorming?

By spontaneously expressing ideas without criticism, this method develops and collects a large number of ideas for a given problem. When brainstorming in a brief meeting, the participating employees encourage each other to come up with new combinations of ideas (group dynamics). This produces more results overall than if everyone were to look for solutions on their own.

How does brainstorming work in a brief meeting?

The brainstorming of a problem must be carried out with the involvement of all employees involved in the sampling process. If the problem is very complex, it should be divided into clearer sub-problems. At the beginning of the brainstorming session, the problem to be addressed is then explained to all those involved in the sampling process. This ensures that everyone has the same understanding of what the problem is. In order to foster mutual encouragement (group dynamic) for new ideas, all suggestions should be written down for everyone to see (e. g. as bullet points or mind maps on a blackboard or whiteboard). It is important here that the project manager acts as the leader or moderator and documents all ideas concerning solutions.

Basic rules for brainstorming

- Express every idea that comes into your head–no matter how nonsensical it may seem. Quantity comes before quality when producing ideas
- No ideas should be criticized or judged (avoid killer phrases in communication)
- Use all contributions as food for thought for further ideas ("That gives me an idea ..."). Change these ideas and combine them with other ideas

What are the benefits of brainstorming?

It is easy to implement without spending a lot of time.
Many possible solutions can be realized in a short time.
Free expression of ideas is encouraged.

What are the disadvantages of brainstorming?

	No visual stimulation (e. g. pictures, sketches) for brainstorming, as the ideas are usually only developed as bullet points.
	The number of solutions or the creativity of the ideas depends solely on the participants.
	If there are a lot of ideas, evaluation can be time-consuming.

10.1.5.2 Mind Mapping

What is mind mapping?

In contrast to brainstorming, a mind map (Figure 10.1) is a graphical representation that combines tasks, words, concepts and ideas with a central main topic. For this purpose, the main topic is written as concisely as possible in the center of a blackboard/whiteboard, a blank sheet of paper, etc. With the help of a mind map, the employees involved in the brainstorming process can set-up an intuitive framework for ideas. This allows long, monotonous bullet points of ideas to be transformed into a colorful, memorable and highly structured presentation. During the brief meeting, the mind map is structured in such a way that each idea is connected to the main topic in the center by a main branch. If further new ideas (sub-ideas) are added to this idea, they are connected as sub-ideas to the existing idea by a "twig". The basic rules here are analogous to brainstorming. It is also important in this method that project manager acts as the leader or moderator and enters all ideas concerning the solution into the mind map.

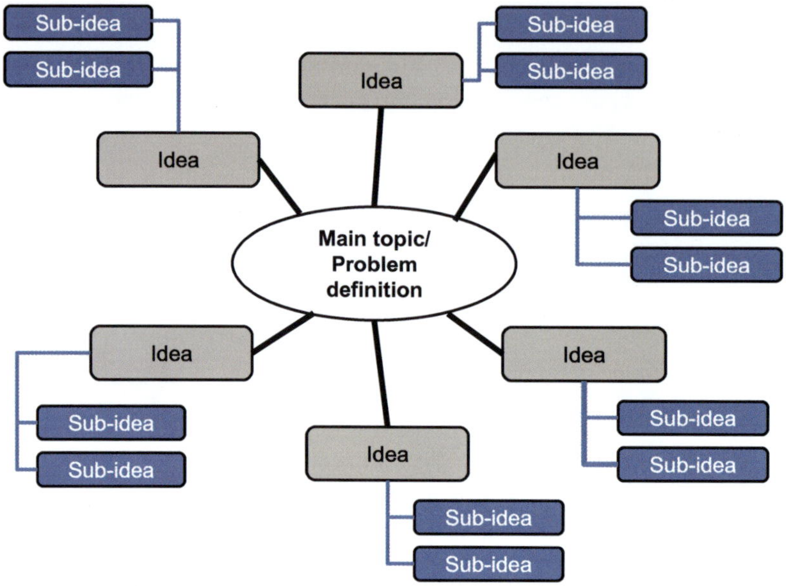

Figure 10.1 Example of a mind map

> **REMEMBER:**
> Mind mapping is a very effective working technique used in a brief meeting to organize thought processes and link images with associations.

How does mind mapping work in a brief meeting?

- **Step 1:** Definition of the main topic or problem
- **Step 2:** A brainstorming session to find ideas and further sub-ideas
- **Step 3:** Creation of a mind map under the direction of the project manager
- **Step 4:** Analysis and decision on actions and next steps

What are the benefits of mind mapping?

👍	The branching can be used to show the relative importance of a thought or an idea for a solution.
👍	The main topic/problem is placed in the center.
👍	Can be combined very well with the brainstorming method.

	Connections quickly become visible.
	Promotes creativity.

What are the disadvantages of mind mapping?

	You have to get used to this format for recording ideas.
	In some cases, low acceptance among people who are not familiar with this creative method.
	At first glance: confusing and convoluted.

10.2 Procedure/Running the Brief Meeting

In the following sections, I will show you how an optimal brief meeting of all employees involved in the process should be organized. To do this, I will discuss the brief meeting step by step and describe all the important tasks of the individual employees involved in the in more detail.

Assessment by the mold sampling specialist

Based on the data documented in the mold sampling report, the "mold sampling specialist" reports on the current problems and deficiencies of the sampling, such as:

- Are there any problems with the filling process for the molded part?
- Are there any problems with demolding of the molded part?
- Is the mold venting adequate?
- Does the injection mold show visible signs of damage?
- Are ejectors, slides/splits, inclined wedges, core pulls, etc. running smoothly?
- Are there any visible defects on the molded part, such as flash formation, ejector marks, scratch marks, white spots, sink marks, blowholes, streaks or interfering weld lines?

- Are the ejectors too far forward or too far back or is there a parting line offset?
- Have the spotting surfaces of the injection mold been checked and are they in order?
- Has anything been damaged in the injection mold during sampling?
- Are there any problems with hot spots on the molded part or the mold?
- Are there any general anomalies in the injection molding process?

Assessment by the mold-making specialist

At the brief meeting, it is very important that all defects and problems in the injection mold identified by the "mold sampling specialist" can be described in precise terms to the "mold-making specialist". All important contributions by the mold maker must be recorded in writing under "Assessment by the mold-making specialist" (Chapter 9, Figure 9.4). The "mold-making specialist" must contribute the following points in the meeting:

- Feedback on the problem solution and the feasibility of changes or corrections to the injection mold or molded part
- Clarification of the time and cost of the required mold modification/correction

Assessment by the design specialist

The "design specialist" acts as a supportive and advisory person for finding solutions to problems. With the help of CAD data, he or she can contribute to problem solutions by checking the feasibility of the injection mold or molded part. If design changes are necessary, all the important details can be discussed and agreed in a brief meeting with all those involved. All important information from the designer's point of view must be entered in the mold sampling report under the section "Assessment by the design specialist" (Chapter 9, Figure 9.3). The general tasks of the "design specialist" in the meeting are as follows:

- Advice on the feasibility of changes or corrections to the injection mold or the molded part
- Clarification of the design time required for changes or corrections to the injection mold or the molded part

Assessment by the quality assurance/metrology specialist

The "quality assurance/metrology specialist" visually assesses the molded part and provides information on the dimensional accuracy and functionality of the molded part. All important points from the specialist's point of view must be documented under “Assessment by the quality control/metrology specialist” (Chapter 9, Figure 9.3). This specialist poses the following questions in the meeting:

- Where are dimensional or optical faults on the molded part?
- Can dimensional problems be solved by tolerance adjustments, displacements?
- Do all the important measured drawing dimensions (such as SC, CC, YS etc.) of the molded parts have a capability (c_{mk} value)?

Assessment by the project manager

The project manager leads the brief meeting and is also responsible for the further documentation of the mold sampling report. He also acts as a moderator in finding solutions using the creative brainstorming and mind mapping methods presented. Once the measures have been defined in common and the next steps discussed, all important facts, such as the duration of the mold correction or changes, the new date for the subsequent sampling, must be clarified and recorded in writing under “Assessment by the project manager” (Chapter 9, Figure 9.4) in the mold sampling report. General tasks in the meeting of the project manager are:

- Management or moderation of the brief meeting
- Moderator role
- Review and approval of the actions taken and the next steps
- Coordinating the definition of actions
- Scheduling of a subsequent mold sampling (iteration loop)

10.3 Definition of Action Plan and Subsequent Procedure

All actions developed in the brief meeting, such as corrections or changes to the injection mold, and subsequent procedure, are documented “Definition of action plan and subsequent procedure” (Chapter 9, Figure 9.5) in the mold sampling report.

10.3.1 Mold Corrections or Changes

The project manager ensures that the sampled injection mold is immediately transferred from the production hall to the tool-making department. There, the tool-making staff can quickly and easily take any necessary corrective measures or changes to the injection mold from the mold sampling report.

NOTE:

While the measures are being taken on the injection mold, the project manager must be updated at agreed intervals using a mold progress plan. This is the only way to plan and initiate further subsequent sampling (iteration loop) in a targeted and timely manner.

Milestone: "Brief meeting and action plan"

- The brief meeting of all employees involved in the mold approval process has been held
- All necessary measures have been taken and recorded in the mold acceptance report
- The injection mold has been handed over to the mold-making department
- Subsequent sampling (iteration loop) by the project manager has been scheduled

11 Subsequent Mold Sampling (Iteration Loop) or Hand-Over to Series Production

Following the brief meeting of all employees involved in the sampling process, the defects in the injection mold or the molded part have been clarified and how these defects are to be rectified have been set out in an action plan. As a result, a subsequent sampling (iteration loop) is carried out on the injection molding machine. However, if there are no defects at the end of the initial sampling, the hand-over process for the injection mold to series production by the "project manager" can take place.

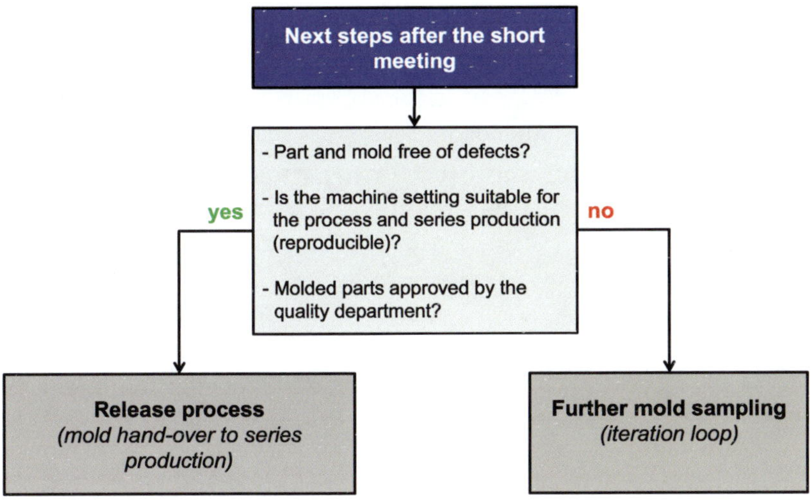

Figure 11.1 Procedure following the brief meeting (schematic)

11.1 Subsequent Mold Sampling (Iteration Loop)

A subsequent mold sampling (iteration loop) is used solely to check whether the defects that occurred in the injection mold or on the molded part could be rectified by the action plan jointly defined in the brief meeting.

11.1.1 Obtaining Information and Preparing for Subsequent Mold Sampling (Iteration Loop)

Each subsequent mold sampling (iteration loop) begins in the same way as the initial sampling, starting with information gathering and preparation for sampling (Chapter 2, Section 2.1 "Information gathering" and 2.2 "Preparation for sampling").

For the employee responsible (e. g. project manager), there are additional points that apply with regard to subsequent mold sampling (iteration loop) which must be observed or checked:

- Have all specified actions as per mold-making been implemented?
- Has the same or an identical injection molding machine from the initial sampling been scheduled for the subsequent sampling?
- Has the "mold sampling specialist" received all the necessary documents?
- Is the old mold report with all sampling documents included in the appendix?
- Has the new mold report been included in the appendix without being filled in, together with all the approval documents?

For the "mold sampling specialist", there are further points that apply with regard to subsequent mold sampling (iteration loop) which must be observed or checked:

- What defects occurred during the initial sampling?
- Which points need to be checked during the subsequent sampling? All defects from the initial sampling are checked again using the sampling checklist
- Are there any special features to look out for in the injection mold?
- All sampling documents that could not be completed during the initial sampling due to defects that occurred in the injection mold or on the molded part must now be documented in the subsequent sampling
- The mold sampling checklist and the mold inspection report must be processed for each subsequent (iteration loop)

11.1.2 Mold Set-Up and Setting the Clamping and Plasticizing Units

In a subsequent mold sampling (iteration loop), the machine setting and process parameters from the machine setting data sheet from the initial sampling are used. This means that all strokes, times, speeds, temperatures and pressures are adopted and set on the injection molding machine. The following important points are to be observed and checked by the "mold sampling specialist":

- Is the injection molding machine the same as or identical with the one used for the initial sampling?
- Is the barrel nozzle bore diameter (mm) the same?
- Have the temperature control circuits on the nozzle and ejector side been connected in the same way using the "Water connection diagram" template from the initial sampling?
- Is the mold temperature control flow rate (1/min) the same?
- Is the melt temperature (°C) the same?

> **NOTE:**
>
> If the initial sampling is aborted early due to very large defects in the injection mold, such as "no demolding", the subsequent sampling (iteration loop) must be carried out in full, starting from Chapter 6 "Filling study, holding pressure and mold clamping force", after the initial sampling has been completed.

11.1.3 Manufacturing Molded Parts and Analysis of the Optimized Basic Setting

If the injection molding machine was set according to the machine setting data sheet from the initial sampling, a mold sampling analysis of the already optimized basic setting must now be commenced, analogous to Section 7.3 "Mold sampling analysis of the basic setting". As this is a subsequent sampling (iteration loop), the analysis only deals specifically with the defects that occurred in the initial sampling and checks whether these could be rectified by the actions defined in the initial sampling.

If all defects have been rectified, all knowledge gained from the subsequent sampling (iteration loop) must be documented in the tool sampling report of the subsequent sampling, as explained in Section 9.3 "Mold sampling report" under "Assessment by the mold sampling specialist".

However, if the same or new defects are present in the injection mold or on the molded part following the mold sampling analysis of the "optimized basic setting", these defects should be examined again in more detail with the help of a subsequent mold sampling analysis (Section 7.3 "Mold sampling analysis of the basic setting"). This has the great advantage that any new knowledge can help to define more targeted measures in the subsequent brief meeting on the subsequent sampling (iteration loop).

11.1.4 Optimization of the "Optimized Basic Setting" for a Subsequent Mold Sampling (Iteration Loop)

A subsequent optimization loop of the machine setting (analogous to Chapter 10 "Brief meeting and definition of action plan") is to be regarded as optional for a subsequent sampling (iteration loop). Optional means that an optimization of the optimized basic setting must be performed in the event of a mold sampling analysis and, as a result, possible new knowledge.

11.1.5 Documentation of Subsequent Mold Sampling (Iteration Loop)

As set out in Chapter 9 "Documentation of the mold sampling", the "mold sampling specialist" documents all important knowledge gained from the subsequent sampling (iteration loop) in the new mold sampling report under the Section "Assessment by the mold sampling specialist". The "mold sampling specialist" informs the "project manager" of the end of the subsequent sampling (iteration loop) on the injection molding machine and hands over the mold sampling report along with all the documents. The "project manager" checks the completeness of the sampling documents again using the checklist at the end. If all the documents for the subsequent sampling (iteration loop) have been filled out completely and correctly, a further brief meeting (ideas-generation meeting) of all employees involved in the sampling process must be convened promptly to draw up the action plan and the next steps to be taken.

11.1.6 Brief Meeting and Action Plan After the Subsequent Sampling (Iteration Loop)

The procedure for a brief meeting of all employees involved in the sampling process must be followed for each subsequent sampling (iteration loop), as well as for a new action plan, as shown in Chapter 10 "Brief meeting and action plan".

11.2 Mold Sampling Cycle

If there are further defects in the injection mold or on the molded part after an initial subsequent inspection (iteration loop), a so-called mold sampling cycle begins (Figure 11.2). This mold sampling cycle in the form of further iteration loops continues until all existing defects in the mold or on the molded part have been eliminated and the injection mold can be handed over for release by the project manager.

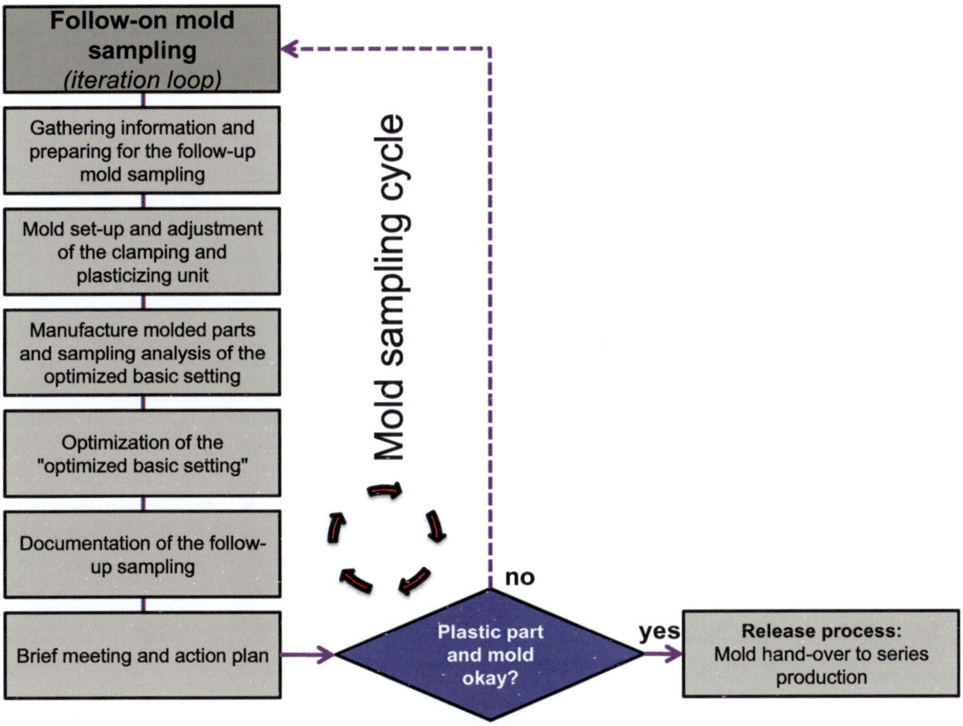

Figure 11.2 Mold sampling cycle

11.3 Hand-over Process (Mold Hand-over to Series Production)

You're done!

You have completed a mold sampling of your new mold. Consequently, at this point:

- The molded part is of good quality
- The mold no longer has any defects
- The machine setting is ready for the process and series production, and
- The molded parts have been approved for all quality characteristics by the person responsible (project manager, quality assurance, customer)

11.3.1 Final Meeting for Mold Hand-over to Series Production

The final step is to use a hand-over process to transfer the injection mold from the sampling phase to series production. Unfortunately, in practice there are often major deficiencies in the hand-over of the mold to series production. Very often the hand-over is characterized by poor communication and documentation, a fact which can lead to considerable problems in the company. One of the deficiencies is that all employees involved in the mold hand-over are not informed at the same time and therefore have a different level of knowledge regarding it. Furthermore, important series processes, such as a series inspection of all quality-relevant features on the molded part, cannot be implemented correctly or even skipped if the hand-over to series production is not official and clearly defined.

In order to ensure an optimal mold hand-over to series production as the last step of the sampling phase, a final meeting of all employees responsible for sampling must be convened together with the responsible employee from series production (foreman, technician, department head).

The aim of a final meeting is, on the one hand, to enable all employees to clarify any outstanding issues with each other and, on the other, to ensure that all employees involved have the same level of knowledge for the hand-over of molds to series production. Furthermore, all the relevant approval documents for series production should be handed over in full with the injection mold. For example, the "quality assurance/ metrology specialist" in the series inspection department knows that the new molded part must be inspected immediately for conformity to the specified quality criteria.

A hand-over protocol (Figure 11.3) should be kept to document the official hand-over of the mold to series production. This protocol serves as later proof that the injection mold was handed over to series production free of defects and with a reliable machine setting.

Protocol for mold hand-over to production					
Article description:			Component:		
Item number:					
Mold number:					
Employees involved in the sign-off at the final meeting:					
Person responsible for sign-off:		Name:	Specialist for construction:		Name:
Machine operator:	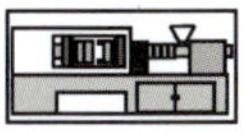	Name:	Specialist for quality management/ measurement technology:		Name:
Specialist for toolmaking:	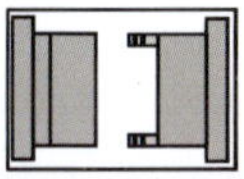	Name:	Series manager:	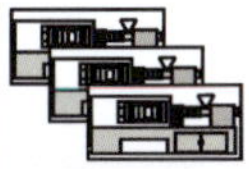	Name:
The following documents have been checked and handed over:					
Machine setting data sheet:				❒ Yes	❒ no
Water connection plan:				❒ Yes	❒ no
Material data sheet on processing instructions from the material manufacturer:				❒ Yes	❒ no
Other:				❒ Yes	❒ no
General/notes/comments:					
Mold hand-over to series production:				❒ Yes	❒ no
Mold hand-over to series production:				❒	❒ no
Person responsible for sign-off:	Signature:		Specialist for construction:		Signature:
Specialist for sampling:	Signature:		Specialist for quality management/ measurement technology:		Signature:
Specialist for sampling:	Signature:		Series manager:		Signature:

Figure 11.3 The "Protocol for mold hand-over to production" template

Procedure at the final meeting for handing over the mold to series production

- **Step 1:** The project manager invites all the sampling employees, including the responsible employee from series production (foreman, technician, department manager)
- **Step 2:** The project manager must prepare the following documents from the last subsequent sampling performed (iteration loop):
  - Machine setting data sheet
  - Material data sheet on processing instructions from the material manufacturer
  - Water connection diagram for mold temperature
- **Step 3:** Hold the final meeting:
  - Hand-over of all important sampling documents
  - Arrangement of mold maintenance intervals
  - Clarification of open points
  - Creation of a mold hand-over protocol to production

Congratulations on successfully completing the **mold sampling phase!**
End of the mold sampling phase and start of series production

12 Redefining the Mold Sampling Process—"Process Engineering Goes Digital"

In this last chapter, I would like to show you how you can implement throughout the company a structured and systematic approach to the entire mold sampling process efficiently and completely digitized and networked.

To this end, I would like to introduce you to a globally unique IT platform for mold sampling and subsequent series processes that I have developed myself: *IMG–Injection Molding*.

But first, let's take a brief look at the world of digitization and Industry 4.0 as well as the needs of and future trends in injection molding technology.

Figure 12.1 Digital mold sampling

12.1 Digitization and Industry 4.0

BACKGROUND KNOWLEDGE:

What does digitization mean ?

The term *digitization* generally refers to the changes to processes, objects and events or the conversion of analogue values into digital formats that are taking place with the increasing use of digital devices.

In the case of mold sampling, this means the storing, distributing or processing of a digital conversion and representation of information and communications in digital form.

The creation, storage, distribution and processing of digital data are carried out using modern information technology, such as computers, smartphones, communications networks, internet applications and databases. The particular advantages of working in digital formats are that they are much more versatile and can process information faster than analogue systems. Especially in the area of mold sampling, this promotes the much-needed communications between individual departments.

What does Industry 4.0 mean?

The underlying idea is to connect people, machines and products directly with each other. For example, screws communicate with assembly robots or self-propelled forklift trucks store goods in high racks. Intelligent machines can also coordinate production processes independently.

Many specialist departments or external companies are involved in the mold sampling process in particular, which contribute step by step to the creation of a molded part or injection molding process. *Digitally networked*, these steps can be better coordinated and the necessary measures can be better planned and implemented.

Use of data: Data on the production process and the condition of a molded part, injection mold or process are brought together and can be evaluated. Joint data utilization and data analysis by the specialist departments provide information on how a product can be manufactured more efficiently. Implementing Industry 4.0 is a complex endeavor. The greater the extent to which processes are digitized and networked, the more interfaces that are created between different players. Uniform norms and standards for plastics processing companies, IT security and data protection play just as central a role as the legal framework, changes in education and work, the development of new business models and the necessary research.

12.2 Why Digitize the Mold Sampling Process?

While the complexity of plastic parts is constantly increasing on one hand, the project phases stretching from the idea to series production are becoming ever shorter on the other. Plastics processing companies have to face this challenge anew every day.

A standardized sampling process plays an important role in the subsequent success of series production, a complex process that involves various specialist departments in the company. All these must work hand in hand at the right time to effectively manage the sampling process. In addition to ever shorter project phases and higher production costs, this task poses technical and logistical problems for many companies.

NOTE:

Most sampling processes today are still managed through a mixture of loose sheets of paper and confusing Excel solutions stored in countless electronic folders (e. g. SharePoint / intranet)!

The basic problem is very often that companies do not perform standardized and digitized mold sampling. As a result, data exchange, communications and documentation before, during and after the mold validation are severely hampered.

The problem is that data have to be taken or compiled from a wide variety of folder structures, e. g. data on the mold, plastic or injection molding machine, even during optimal preparation for sampling. Furthermore, important information, such as the mold temperature control plan or data on the hot runner system, are usually only available, if at all, as loose sheets of paper at the start of sampling at the machine. There is simply no IT system that contains and retains all the information for a molded part over the entire project duration.

12.2.1 Intention Behind the "IMG" IT Platform

Figure 12.2 The IMG platform: Process engineering goes digital

Many years of experience in successfully introducing standardized process standards in companies have shown that more and more computer-aided systems, such as ERP, PDA or machine production control stations, are being introduced. Unfortunately, the most important process of an injection molding company, mold sampling, is often worked around, but not directly addressed. Even in modern companies, this continues to lead to less-than-optimal mold sampling processes and disrupts efficiency, targeted definition of action plans and communications and documentation between the individual departments.

For this reason, the IT platform "IMG–Injection Molding Guide" has been developed and perfected from practice for use in practice as a way of solving these problems in a company quickly and easily.

The world's first IT platform for mold sampling and series processes offers impressive technical advantages for the resulting injection molding process that cannot be achieved with the conventional everyday measures employed in a company. The operator's technical work becomes standardized, digitized, networked and consequently implemented more sustainably for the injection molding company.

12.2.2 Benefits of the IMG Platform for the Company

Turnkey standardized process

- Problem, time and cost reductions across the entire sampling process
  - Reduction of throughput times up to series production
  - Conservation of resources
  - Faster implementation of a single mold sampling
- Improved communications and documentation between the specialist departments
- Improved productivity, cost-effectiveness and energy efficiency of every injection molding process
- Clear, transparent and structured process

Digitized process in a single, compact IT platform

- Faster exchange of information and data transfer before, during and after mold sampling
- Parallel working of all departments involved in the project before, during and after mold sampling saves resources, time and money
- Intelligent document management
- Use of tablet, laptop etc to perform the implement the sampling process at external suppliers (e. g. initial mold commissioning at the external mold maker)

Promotion of basic and advanced training through integrated help

- Every employee works at the same necessary high level, regardless of whether they are skilled or semi-skilled
- Guide functions support every phase of the molding process with detailed instructions and assistance
- Large database contains strategies for eliminating optical and thermal molding defects by means of the integrated “IMD–Injection Molding Doctor®”
- Integrated knowledge management
- Content serve as internal teaching material
- Elimination of external cost-intensive training courses
- Rapid familiarization of new skilled or semi-skilled workers

12.3 Brief Insights into the IMG

The IMG enables a process-oriented and user-friendly standardized, digitized and networked platform within the company that can be used by all departments, such as project managers, developers, designers, quality assurance, mold-making and production. As a workflow-supported control system, the platform also offers a digital transformation of your entire process engineering (process engineering goes digital) (Figure 12.3).

Information on plastics, machines, peripherals, process data and much more is stored in an SQL database. The fact that the data can be retrieved at any time means that sampling processes can be implemented faster, and outlay on costs and time are reduced.

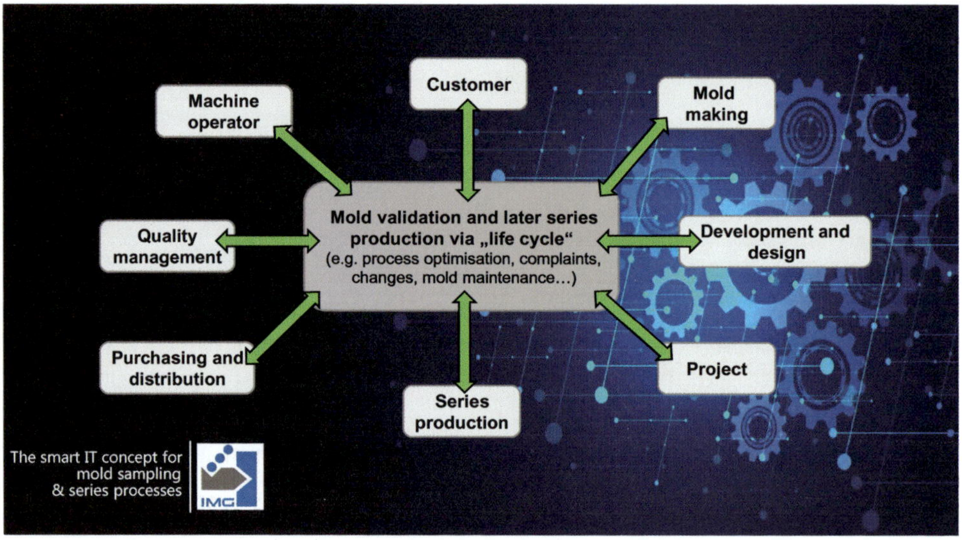

Figure 12.3 IT control system and active work platform for all specialist departments

12.3.1 Insights into Set-Up, Content and Functions

Master data

The master data (Figure 12.4) help to save time resources in advance.

The IMG offers the following advantages:

- All necessary data can be stored quickly and easily and serve as information and can be used for data transfer for the various departments throughout the sampling process.

- The internal database can be used to save all the data created, such as data on the plastic to be processed or the injection molding machine used.
- All data formats, documents, templates, photos etc. can be integrated and edited.
- Numerous links from the master data contribute to more effective, time-saving processing of the various sampling steps in the area of process engineering later on the IMG platform.

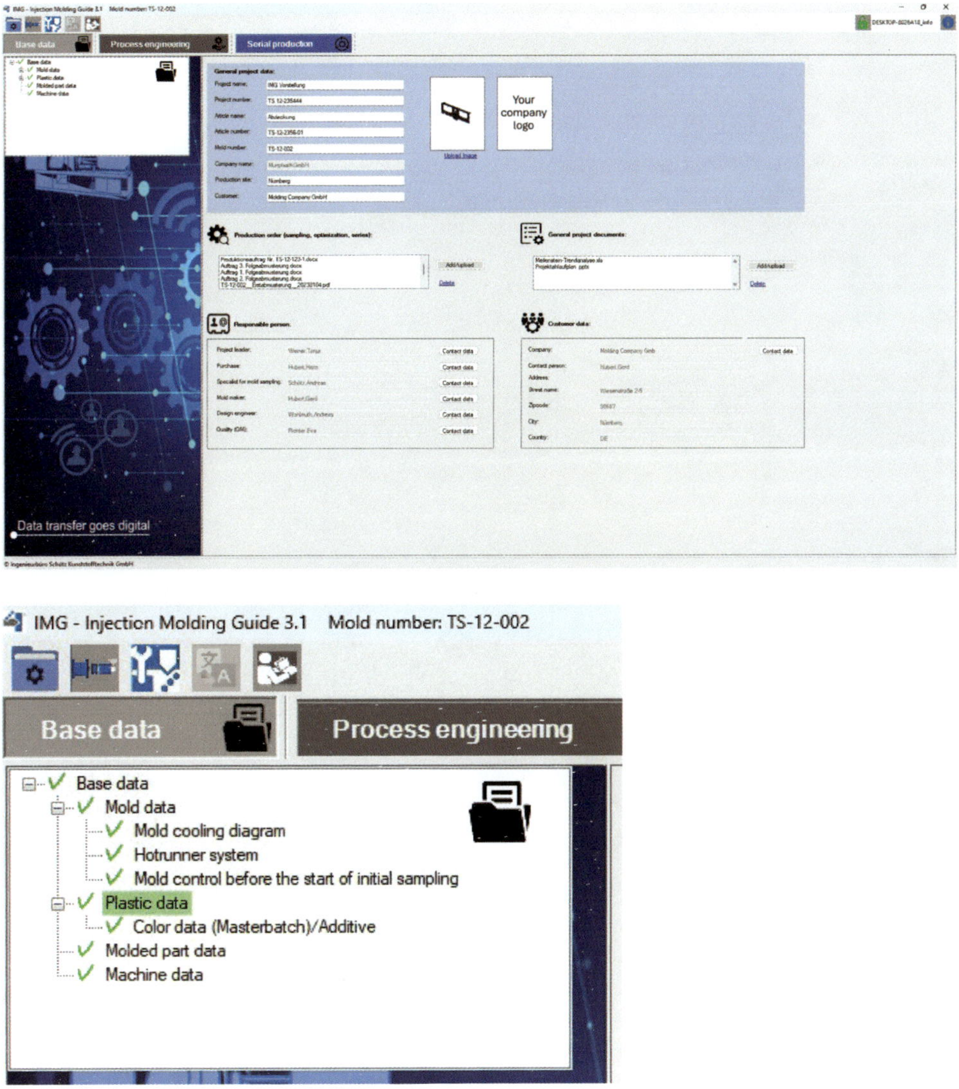

Figure 12.4 Excerpts from master data

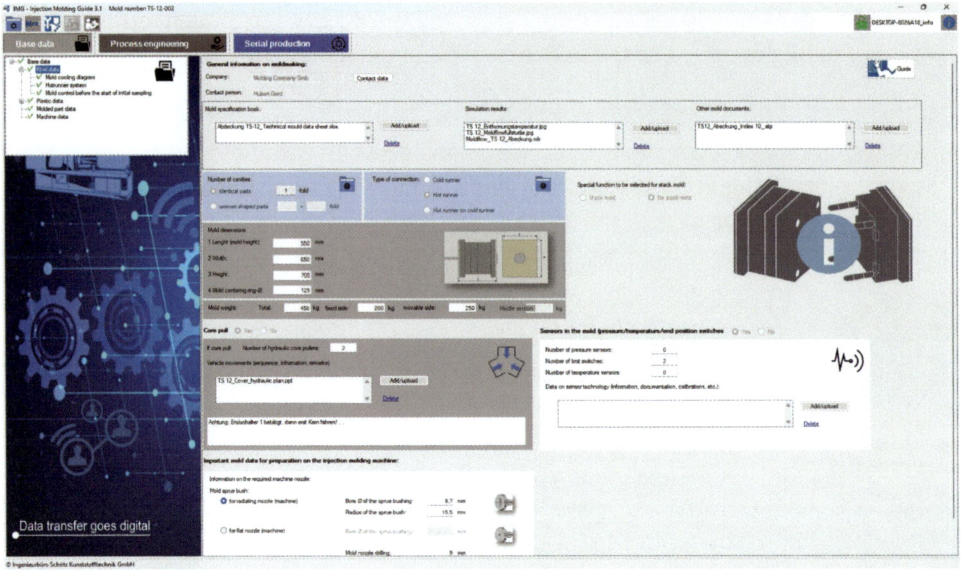

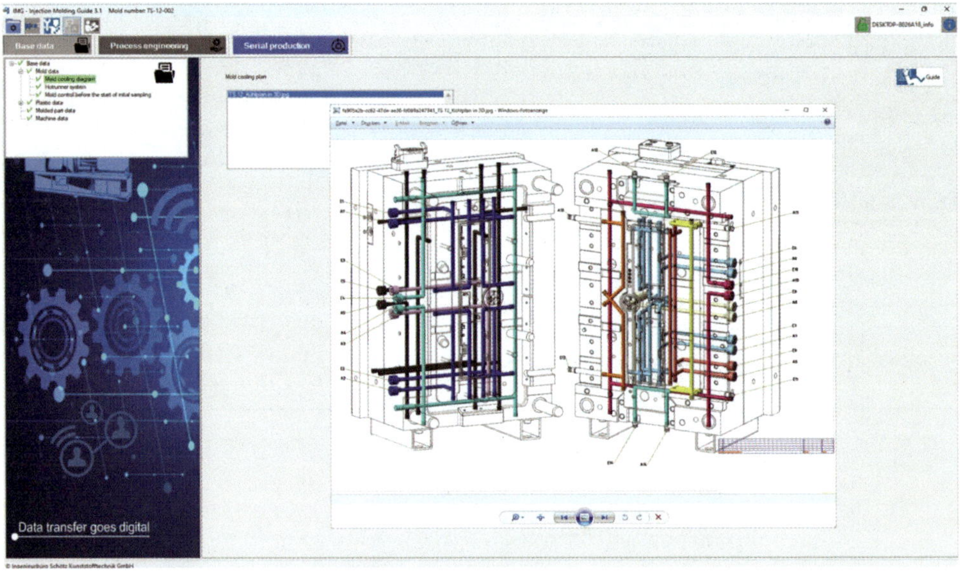

Figure 12.4 Excerpts from master data (*continued*)

Operational implementation continues under process engineering

Figure 12.5 Process engineering

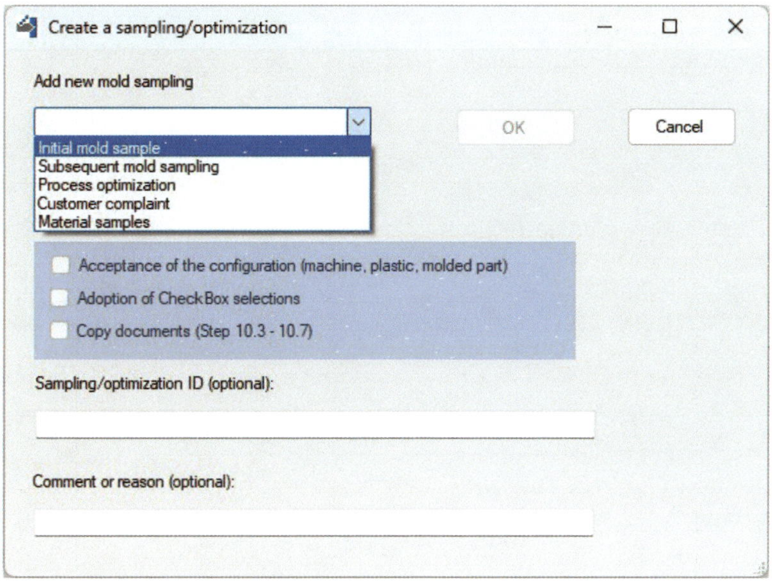

Figure 12.6 Add new mold sampling

Once all the important background information has been compiled, inserted and saved under master data, the operational part on the injection molding machine follows. The user can select from the following categories under "Add new mold sampling" (Figure 12.6).

The user (Figure 12.7) then uses the workflow-supported IT control system in a structured and analytical manner to work their way through the entire sampling/process optimization on the machine.

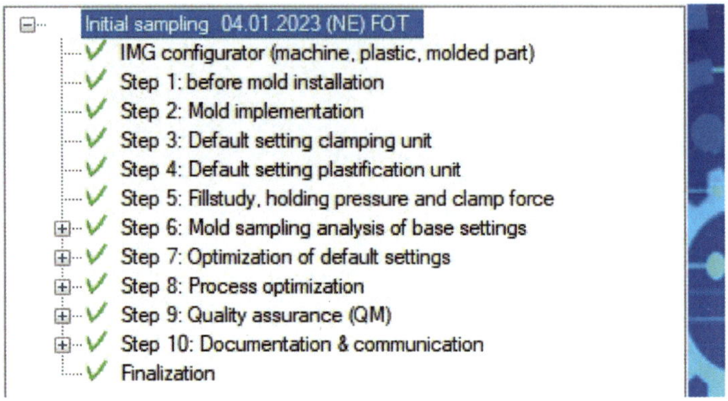

Figure 12.7 Tree structure

All important operational sampling content can be implemented in ten steps with the help of targeted checklists (queries always occur at the right time), documentation options, templates and integrated calculation operations (Figure 12.8).

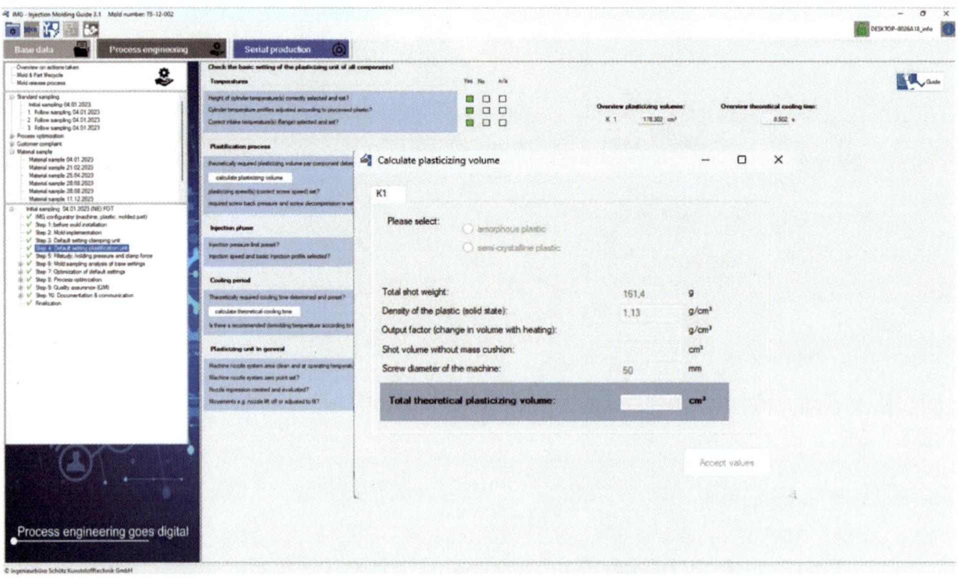

Figure 12.8 Extract from the operational section under process engineering

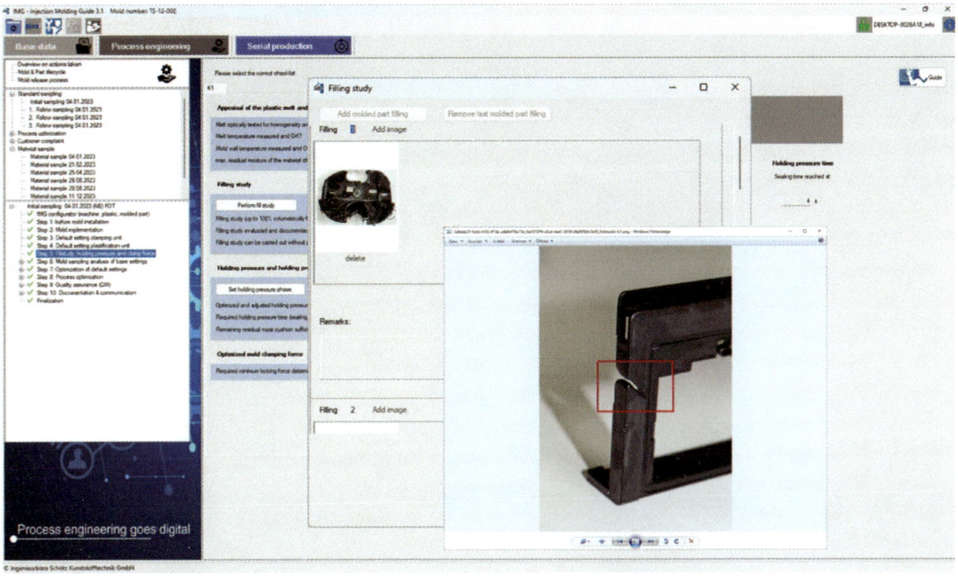

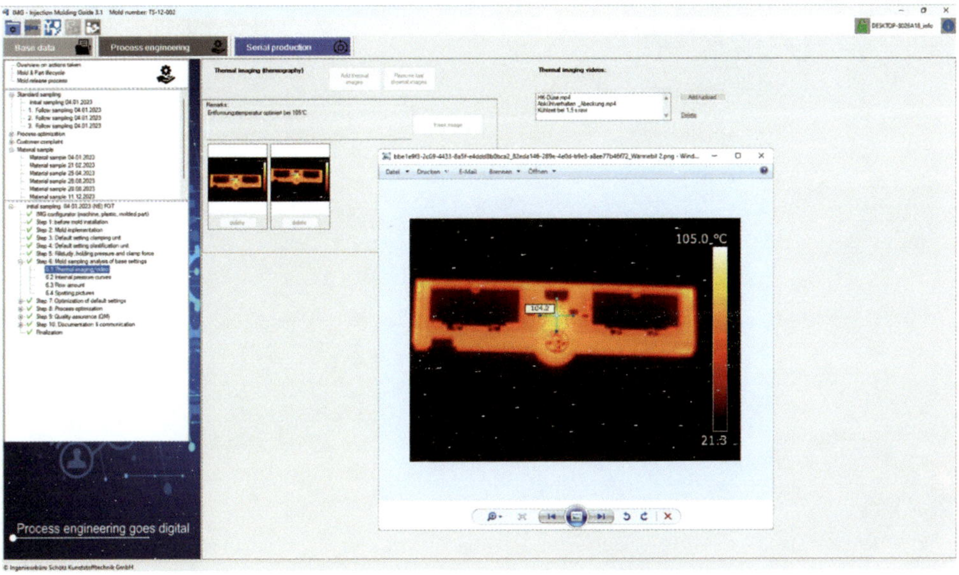

Figure 12.8 Extract from the operational section under process engineering (*continued*)

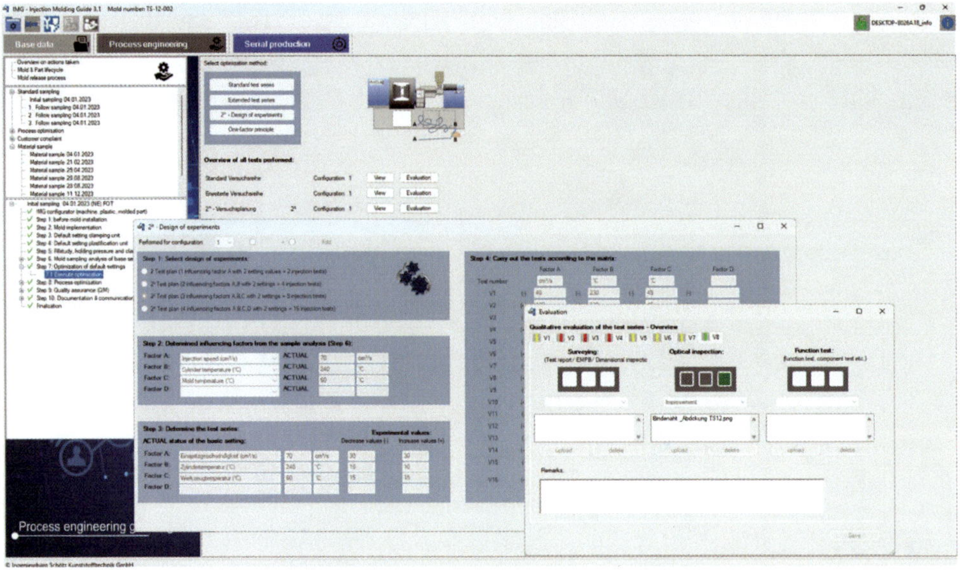

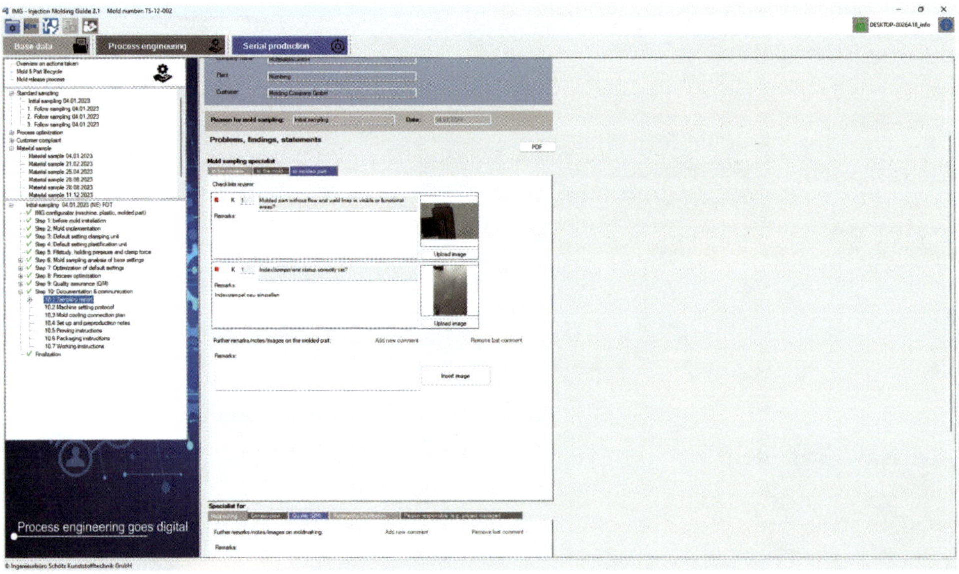

Figure 12.8 Extract from the operational section under process engineering (*continued*)

12.3.2 About the IMG Platform

The IMG saves all integrated data and documents on the internally protected database. This has the great advantage that the internal and externally protected database can be accessed quickly for further subsequent sampling or optimizations.

In addition to the predefined contents of the checklists, it is also possible to add and remove your own points, meaning that the queries can be customized even more individually to the company using checklists.

Another special feature of the software is the integrated "checklist review", which automatically stores and saves all items check-marked with "*No*" in the sampling report during operational implementation.

All existing document templates, e. g. from quality assurance, can be easily integrated and edited in IMG and are automatically saved in the internal database.

Selected documents can be converted from the IMG into PDF format for internal use or for customers and then displayed. These documents must of course be processed as controlled documents in accordance with certification and audit requirements.

Included: Guide function

If the user requires more background information during running sampling or process optimization, e. g. about the setting strategy for the basic machine setting and process parameters, the integrated guide contains numerous instructions, graphics, tables, photos/images for performing all sampling or optimization steps (Figure 12.9).

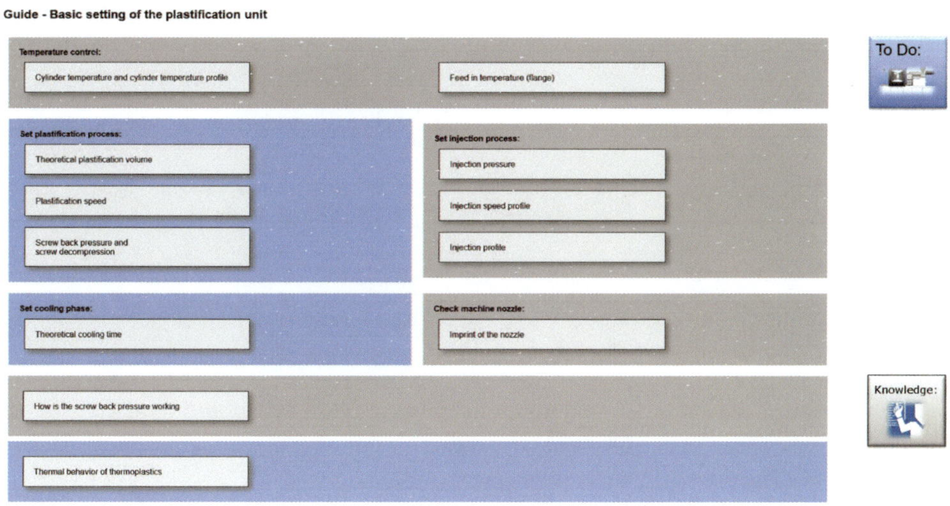

Figure 12.9 Excerpts from the IMG guide

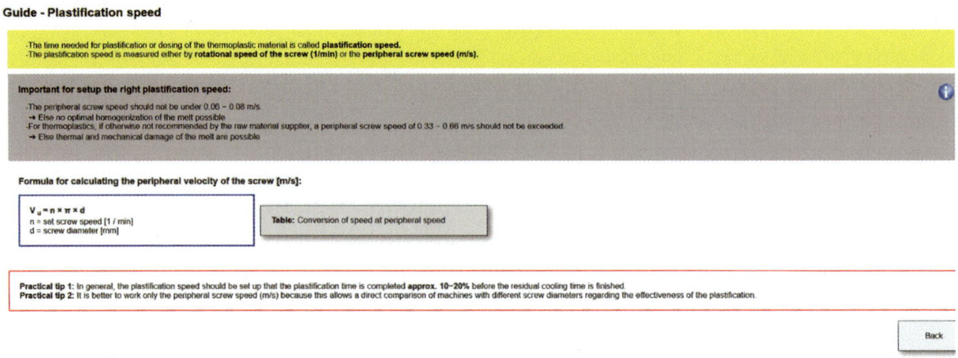

Figure 12.9 Excerpts from the IMG guide (*continued*)

Included: Injection Molding Doctor®

The Injection Molding Doctor® (IMD) offers extensive support for error detection and a targeted action plan for optimization.

The IMD differs from all other defect catalogues and action plans already available on the market, whether in printed form or as an app, on account of the following features:

- In addition to the classic optical fault patterns, possible thermal sources of faults are covered in detail using thermal images
- The user is given guidance in the form of three practical, fast and effective steps on the machine
- Additional info buttons for targeted root-cause analysis and troubleshooting provide further background information, recommendations and more precise assistance
- You can add and save your own photos for each fault pattern in the "Image catalogue"
- Texts, explanations and images/graphics are kept short and easy to understand
- Thanks to its didactic set-up, the IMD can also be used specifically for employee training on fault patterns that occur during injection molding

Each employee can perform various tasks, ranging from systematic and structured defect detection (Figure 12.10) through to action plan and optimization strategies in three fast and effective steps directly at the machine.

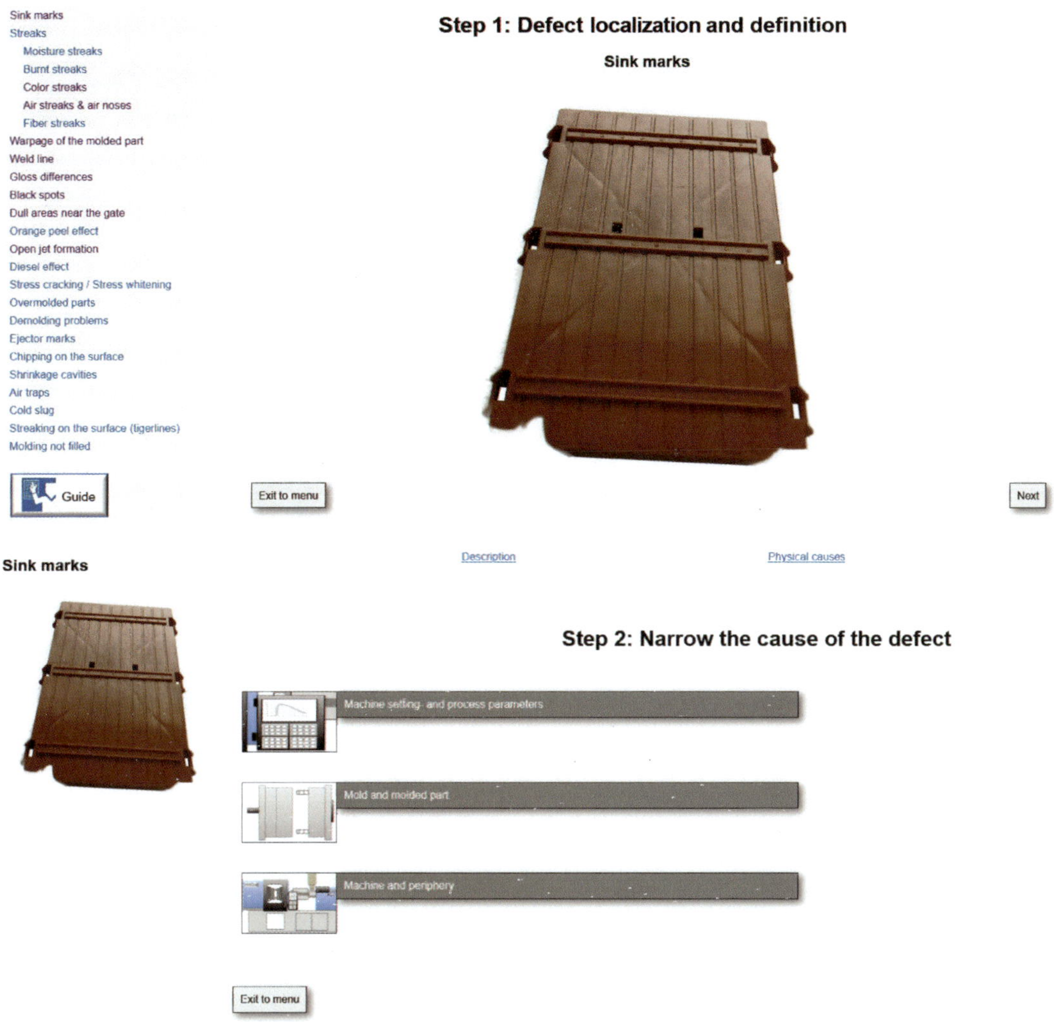

Figure 12.10 Excerpts from the **Injection Molding Doctor®**

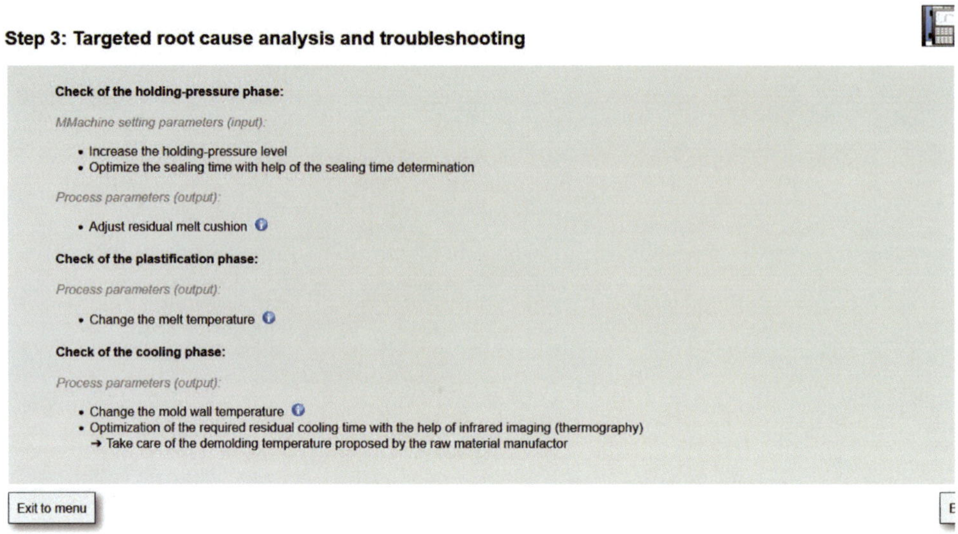

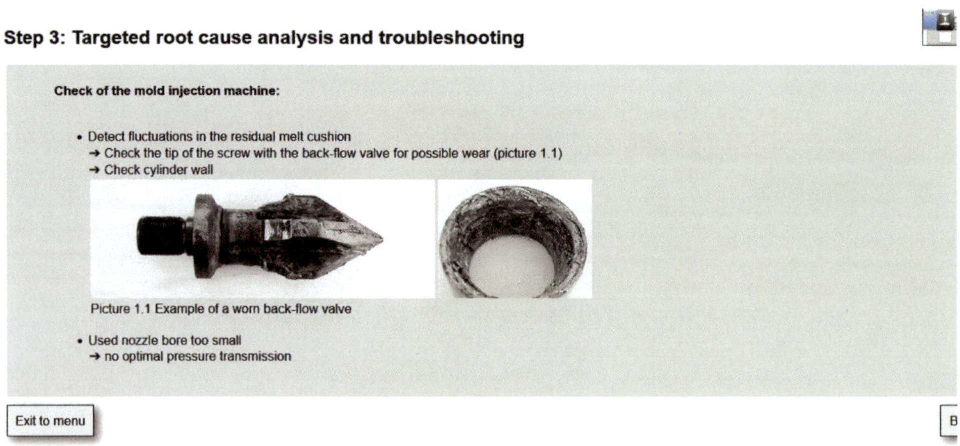

Figure 12.10 Excerpts from the **Injection Molding Doctor®** (*continued*)

Another highlight of the Injection Molding Doctor® is an integrated info button for selected actions. This leads the user to further background information and possible troubleshooting strategies.

12.4 Using the IMG Platform Beyond the Production Life Cycle

Figure 12.11 Excerpt from series production

At the beginning of the book, we talked a lot about all the problems involved in a sampling process. However, the issue of "How do I safeguard my ongoing series process?" is usually a problem that is not optimally solved within the company.

You should ask the following questions critically and self-reflectively within the company:

- Is there only ever one released master setting per article, which is fully validated and linked to the current release status?
- How does the worker/setter know which settings are on the current release status?
- How do you secure the master setting in the ongoing series process?
- How do you ensure 100% transparency, communications and documentation when intervening in the ongoing series process?
- Do you still have process folders in paper form at the injection molding machine?
- Are these folders containing the released data, figures and facts always up to date in the production shop?
- Is the employee at the machine actively supported in the event of problems in the ongoing series process?

- What form of documentation & means of communication with other departments does the employee at the machine have for active troubleshooting?
- Are there digital solution strategies for the employees at the machine when problems or malfunctions occur in the ongoing series process?
- Do you have an internal process standard for "Troubleshooting and intervention in the ongoing series process"?

NOTE:

The process engineering and its challenges in everyday life do not end with the sampling process. Continue to utilize all the benefits of IMG beyond the production life cycle.

12.4.1 A Short Overview of Series Production

If the above questions are not optimally resolved in the company or are left mostly untouched, the effectiveness of the previous sampling phase is unfortunately undermined. This means that your goal should also be a transparent, systematic and analytical approach to "interventions in the ongoing series process".

12.4.1.1 Digital Process Folder

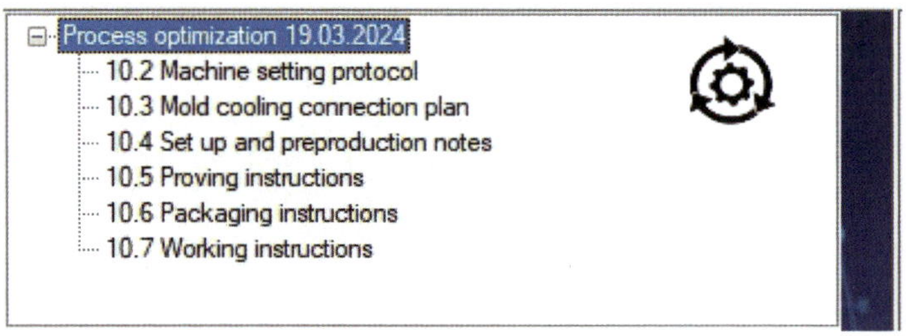

Figure 12.12 Digital process folder

- Paper no longer used at the machine
- Automatic update after each release process
- Backup of the master setting (data set) & all relevant production documents
- High process reliability and fast, secure data transfer at the machine

12.4.1.2 Digital Fault Catalogue

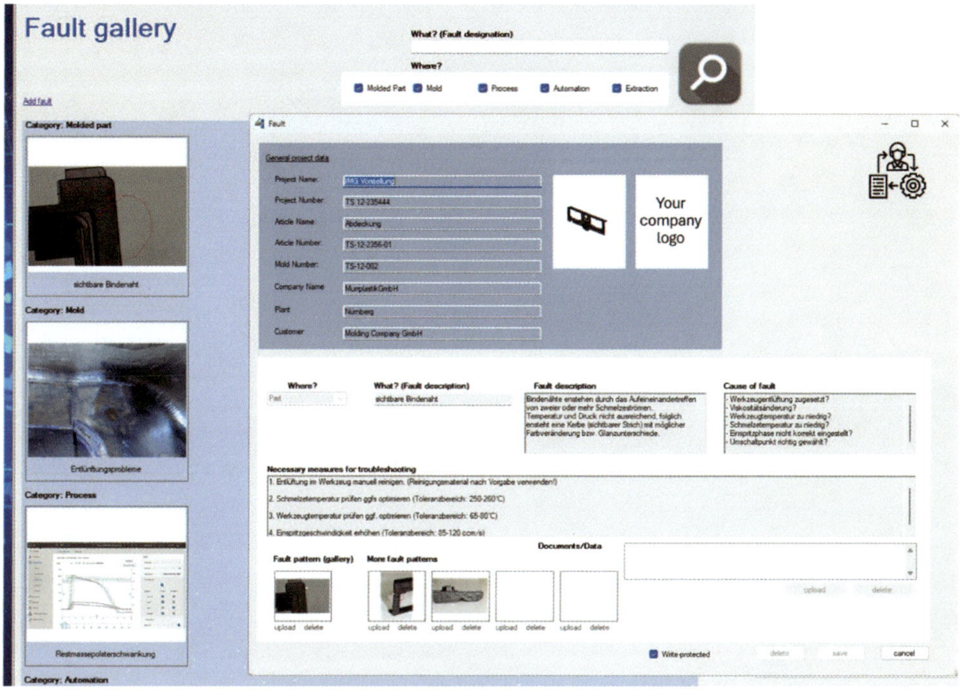

Figure 12.13 Excerpt from the fault catalogue

- Creation of your own customized fault database
- Intelligent search functions for fast queries in the ongoing series process

12.4.1.3 Digital Process History

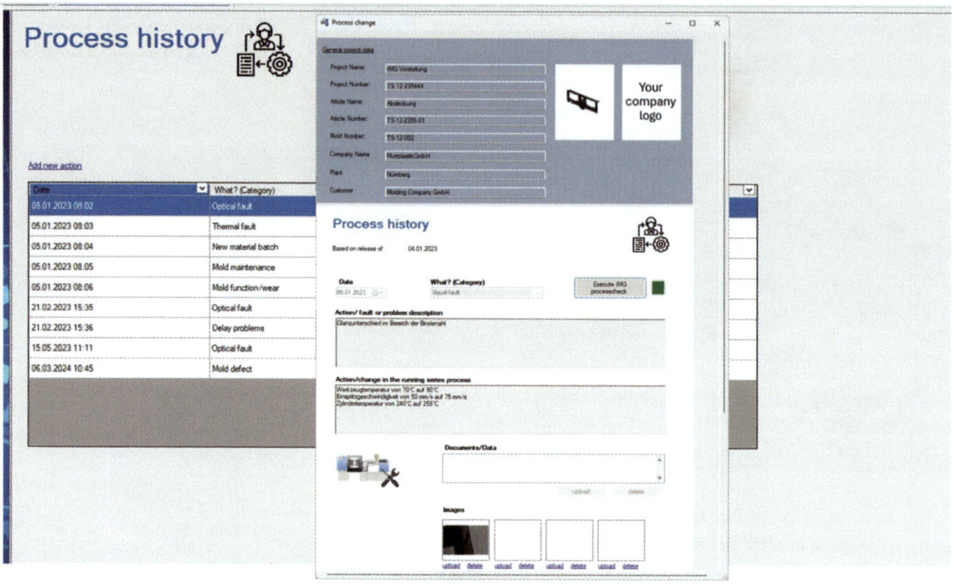

Figure 12.14 Excerpt from the process history

- Employee guided by "IMG process check" through the procedure for dealing with faults in the ongoing series process
- Simple and intuitive user masks
- Rapid transfer of information to employees responsible for the process

12.4.1.4 Smart Digital Mold Maintenance

The new IMG platform, release 3.0 and above, includes smart tool management with a maintenance tool for your injection molds and a digital tool history.

In addition to functionalities, such as automatic information flow by e-mail and a clear matrix (project manager) for all molds, many other new features have been integrated.

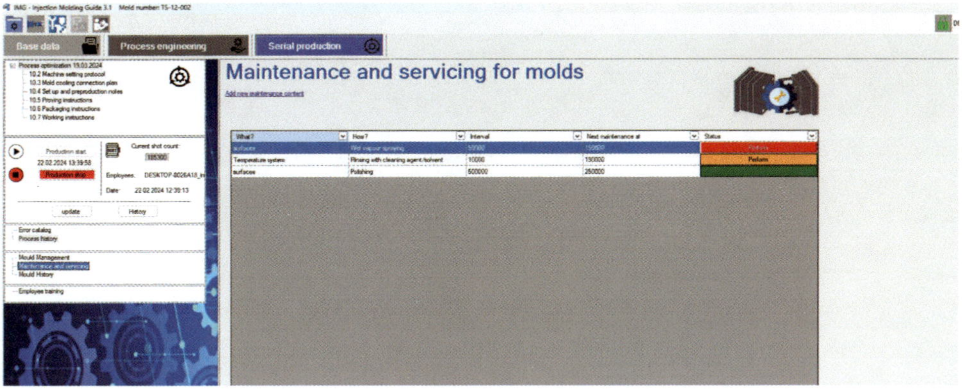

Figure 12.15 Excerpt from mold management

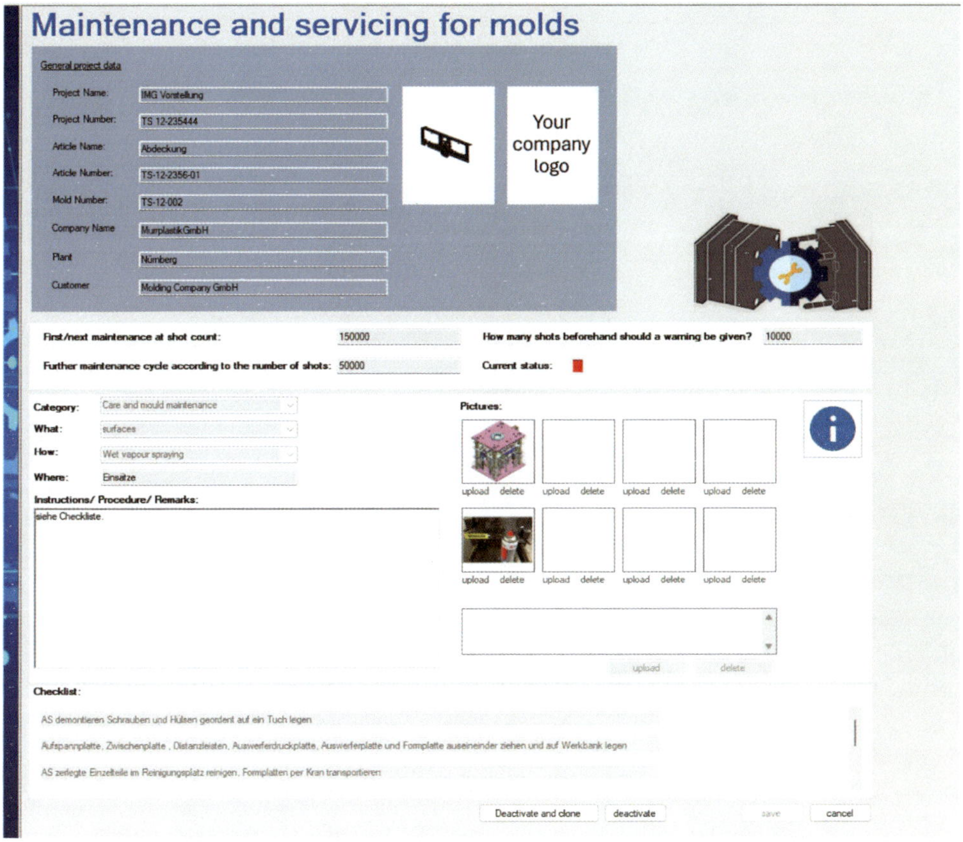

Figure 12.16 Excerpt from the mold history

Figure 12.16 Excerpt from the mold history (*continued*)

12.4.2 Summary

The IMG platform is a toolbox that allows you to quickly and easily implement in full everything that has been discussed in this book, from information gathering and planning, to the operational part on the machine, to the definition of the action plan through to hand-over to series production.

13 Final Words

This text book on mold sampling attempts to highlight the need for mold sampling. It also shows how this complex sampling process should be structured, analyzed, digitized and networked in every company.

In keeping with the motto "*standardize, digitize and network one of the most important processes in your company: the holistic mold sampling phase and process optimization*", I would like to take this opportunity to wish you every success with the implementation.

The author, July 2024

Andreas Schötz

Index